Matter, Dark Matter, and Anti-Matter

In Search of the Hidden Universe

Alain Mazure and Vincent Le Brun

Matter, Dark Matter, and Anti-Matter

In Search of the Hidden Universe

Dr Alain Mazure
Director of Research
Laboratoire d'Astrophysique
de Marseille
CNRS
France

Dr Vincent Le Brun
Assistant Professor
Laboratoire d'Astrophysique
de Marseille
Aix-Marseille Université
France

Original French edition: *Matières et antimatière – A la recherché de la matière perdue*
Published © Editions Dunod, Paris 2009
Ouvrage publié avec le concours du Ministère français chargé de la Culture - Centre National du Livre
This work has been published with the financial help of the French Ministère de la Culture - Centre National du Livre

Translator: Bob Mizon, 38 The Vineries, Colehill, Wimborne, Dorset, UK

SPRINGER–PRAXIS BOOKS IN POPULAR ASTRONOMY
SUBJECT *ADVISORY EDITOR*: John Mason, M.B.E., B.Sc., M.Sc., Ph.D.

ISBN 978-1-4419-8821-8 e-ISBN 978-1-4419-8822-5
DOI 10.1007/978-1-4419-8822-5
Springer New York Dordrecht Heidelberg London

Library of Congress Control Number: 2011940316

Cover design: Jim Wilkie
Translation editor: Dr John Mason
Typesetting: BookEns, Royston, Herts., UK

Printed on acid-free paper

Springer is part of Springer Science+Business Media (www.springer.com)

To our families

For all the nights spent on telescopes
and all the days on computers. . . .

Contents

Foreword

An absolute triumph of modern astrophysics was the precise measurement of the content of our universe in ordinary matter. Scientists refer to this component as baryons, of which the most familiar are the protons and neutrons which make up most of the mass in the atoms and molecules of the universe.

Around the turn of the past century, researchers executed two entirely independent experiments to 'weigh' the baryons in our universe. In one experiment, several teams used balloon and space-borne telescopes to measure tiny oscillations in the relic radiation field of the Big Bang (termed the cosmic microwave background or CMB). The relative amplitudes of these CMB waves are sensitive to the amount of baryons in our universe and thereby enabled the baryon content to be computed.

In the other experiment, astronomers counted the relative abundance of two species, hydrogen and its heavier isotope deuterium, by analyzing spectra of some of the most distant and exotic objects in the universe (quasars). Their analysis set an entirely independent constraint on the density of baryonic matter. Each of these experiments had its own set of technical challenges and uncertainties, and each required complex theoretical calculations of the universe at very early times. Amazingly, the two experiments agreed – to very high precision – that just 4 percent of the mass and energy of our universe exists in the form of baryons.

The profound implications of this measurement are twofold: First, it firmly established that the overwhelming majority of mass and energy in the universe is in a form altogether alien to our everyday experience. We refer to these other components as 'dark matter' and 'dark energy'. Indeed, exploration of these dark components – the remaining 96% of our universe – is a primary focus of current research.

Second, the measurement afforded scientists the opportunity to compare a census of baryons in the modern universe with the estimated total. To our astonishment, the first census (performed by a pair of astrophysicists at Princeton University) revealed a stunning puzzle: luminous matter in stars and galaxies, and hot gas detected in galaxy clusters, and intergalactic hydrogen accounts for only about half of the total baryons in our universe. In essence, half of the baryons in the present day universe are 'missing'!

This realization inspired research groups from across the world (including my own) to begin a dedicated search for these missing baryons. Our efforts have focused primarily on obtaining high-precision spectroscopy at ultraviolet wavelengths of distant quasars (the highly luminous cores of galaxies, powered by supermassive black holes) using the Far-Ultraviolet Space Explorer and Hubble Space Telescope. These data permit the search for baryons in a warm and diffuse phase that generally precludes easy detection. Despite some success, the search for the missing baryons continues. This quest melds careful observational experiment using the world's most powerful telescopes with dedicated theoretical calculations on the fastest super-computers.

In this book, Alain Mazure and Vincent Le Brun offer a fascinating review of the history of the baryonic measurements, the discovery of the missing baryons problem, the techniques used to perform a census in the nearby and distant universe, and the theoretical efforts that guide this research. Their book marks the first complete account of our understanding of ordinary matter. It is beautifully illustrated and written with simple yet physically insightful terms. I will recommend it not only to friends and family with a scientific interest, but also to first-year PhD students who intend to carry out research this field.

J. Xavier Prochaska
Professor of Astronomy & Astrophysics
University of California, Santa Cruz
Astronomer, University of California Observatories

Authors' Preface to English Language Edition

For both of us, this book is the result of a journey through the still open questions raised by the cosmological model. We were, of course, confronted with the dark matter and dark energy enigma whose elucidation calls for 'new physics' and the elaboration of new instrumentation, both requiring considerable efforts on the part of astrophysicists and particle physicists.

However, apart the exploration of this 'dark' sector, it appears that a 'tiny' problem is still unresolved in the 'ordinary' and in principle well known 'bright' sector: baryons seem missing in our immediate environment!

The fact that a fraction of the ordinary matter, which appears as only a slight 'froth' above a deep ocean of dark matter and dark energy, is lacking could seem unimportant. But firstly, as human beings we must be concerned with it since we, and the planet we live on, are made from this froth! And there is yet another reason to hunt for this missing part. It is always a duty of scientists to check the global coherence of a model and confirm that there is no inconsistency which could reveal some fundamental failure.

So, the hunt for the missing baryons is on!

Alain Mazure and Vincent Le Brun
Laboratoire d'Astrophysique de Marseille
June 2011

Acknowledgments

We would first like to thank both Praxis Publishing and Springer who have made possible this English language translation from the original French edition, and the final publication of this book.

We also would like to thank our colleagues in Marseille, both astrophysicists and particle physicists, for all the debates on dark energy, dark matter, reionization, primeval stars and galaxies, and missing baryons. This book is, in part, the fruit of those animated discussions.

Illustrations

Introduction
An intriguing absence

Absence of proof is not proof of absence

For more than a decade now, the dark side of the universe has been headline news. Detailed studies of the rotation of spiral galaxies, and 'mirages' created by clusters of galaxies bending the light from very remote objects, have convinced astronomers of the presence of impressive quantities of *dark matter* in the cosmos. Moreover, in the 1990s, it was discovered that the expansion of the universe, far from slowing down, as predicted in the standard cosmological model, has recently (on the cosmological scale, that is: some four to five billion years ago) entered a phase of acceleration. This discovery

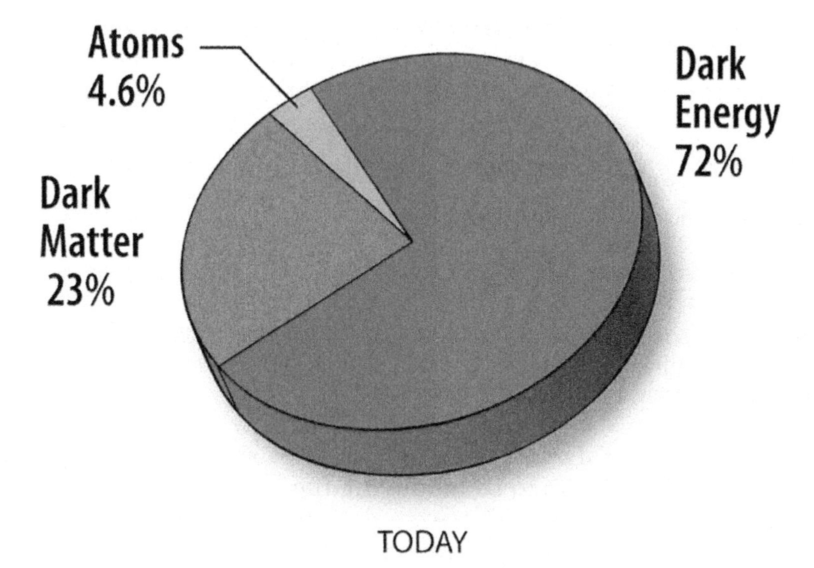

Figure 0.1 Astronomers' estimates of the energy/matter content of the universe lead to the surprising conclusion that ordinary ('baryonic') matter, essentially responsible for the radiation emitted by stars and galaxies, represents but a small fraction ($\sim$4–5%) of the global contents of the cosmos, which is dominated by dark matter and dark energy. Dark matter does not emit or absorb light; it has only been detected indirectly by its gravity. Dark energy acts as a sort of an 'anti-gravity'. This energy, distinct from dark matter, is thought to be responsible for the present-day acceleration of the universal expansion.

Periodic Table of the Elements

	IA	IIA	IIIB	IVB	VB	VIB	VIIB	VII	VII	VII	IB	IIB	IIIA	IVA	VA	VIA	VIIA	0
1	1 H																	2 He
2	3 Li	4 Be											5 B	6 C	7 N	8 O	9 F	10 Ne
3	11 Na	11 Mg											13 Al	14 Si	15 P	16 S	17 Cl	18 Ar
4	19 K	20 Ca	21 Sc	22 Ti	23 V	24 Cr	25 Mn	26 Fe	27 Co	28 Ni	29 Cu	30 Zn	31 Ga	32 Ge	33 As	34 Se	35 Br	36 Kr
5	37 Rb	38 Sr	39 Y	40 Zr	41 Nb	42 Mo	43 Tc	44 Ru	45 Rh	46 Pd	47 Ag	48 Cd	49 In	50 Sn	51 Sb	52 Te	53 I	54 Xe
6	55 Cs	56 Ba	57 *La	72 Hf	73 Ta	74 W	75 Re	76 Os	77 Ir	78 Pt	79 Au	80 Hg	81 Tl	82 Pb	83 Bi	84 Po	85 At	86 Rn
7	87 Fr	88 Ra	89 +Ac	104 Rf	105 Ha	106 Sg	107 Ns	108 Hs	109 Mt	110 110	111 111	112 112	113 113					

* Lanthanide Series	58 Ce	59 Pr	60 Nd	61 Pm	62 Sm	63 Eu	64 Gd	65 Tb	66 Dy	67 Ho	68 Er	69 Tm	70 Yb	71 Lu
+ Actinide Series	90 Th	91 Pa	92 U	93 Np	94 Pu	95 Am	96 Cm	97 Bk	98 Cf	99 Es	100 Fm	101 Md	102 No	103 Lr

Families of Elements

Each element in the periodic table has distinctive properties.
When elements have similar properties they are grouped into families.

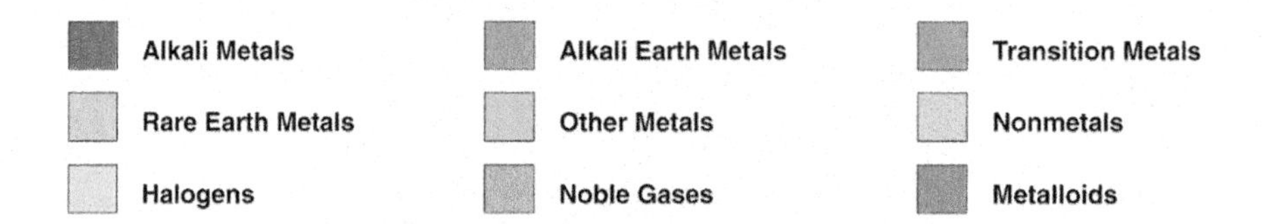

Figure 0.2 A modern version of Dmitri Mendeleev's Periodic Table of the Elements. Each element in the table has distinctive properties; when elements have similar properties they are grouped into families.

implies the existence of *dark energy*. The nature of these 'dark' ingredients remains a mystery. Ought we to be modifying the laws of gravity in order to make sense of these phenomena? Are we dealing here with particles as yet undetectable, but predicted by certain extensions of the standard model of particle physics? Should we be looking for our answer towards the quantum vacuum energy of the primordial universe, or some similar idea? The most striking thing is that these two exotic associates, objects of fervent scrutiny by astronomers and physicists, seem to comprise about 95 percent of the matter/energy content of the universe (Figure 0.1).

So, what of ordinary matter? Although we are permanently immersed in a sea of dark particles, primordial neutrinos and photons from 'fossil' cosmological radiation, both we and our environment are made of ordinary, baryonic matter, as depicted in a famous illustration: the periodic table of Dmitri Mendeleev (Figure 0.2). Even if only 15-20 percent of matter is baryonic matter, or if it represents only 4-5 percent of the total content of the universe, a coherent estimate of its contribution throughout cosmic history can only strengthen the underlying cosmological model.

In the context of the Big Bang model, and with the aid of nuclear physics, we can understand the origin of these baryons (atomic protons and neutrons) and accurately estimate their initial contribution to the contents of the cosmos. However, if we track them through the course of the last fourteen billion years, it seems that they are in short supply in the present epoch. Now, when we observe the fossil cosmic background radiation from the era of the recombination of the universe, 300,000 years after the Big Bang, and the harmonics in associated cosmic waves, we can infer that, at that epoch, the baryon count was holding its own. Also, the exploration of the 'Lyman-alpha forest', which gives us information about regions of diffuse gas throughout the cosmos when it was two billion years old, reveals that the abundance of baryons was then equal to its primordial value.

So where have the baryons now gone? They cannot simply have disappeared. It is the story of this ordinary matter, and the quest for what are known as the 'missing' baryons, that we shall recount here. We shall present the inventory of matter, examining it in all its forms and enquiring by the way into the fate of its twin, anti-matter. We shall show how technological progress has constantly accompanied this research, in tandem with the evolution of ideas, and how the combined effect of these advances might help us definitively to lift the veil from these hidden baryons. Finally, we bring in the close links between matter and types of energy, cosmology and particles, the macrocosm and the microcosm.

1 Matter concentrated

"Too much abundance impoverishes the material." Boileau

We shall set off in pursuit of so-called ordinary or baryonic matter, i.e. the protons and neutrons gathered together in the nuclei of the atoms that constitute all the elements to be found on Earth and in the universe. However, throughout what is termed the thermal history[1] of the universe, these baryons have been subject to transformations. In particular, they can remain as atoms or regroup themselves into molecules, and may or may not undergo considerable variations in temperature according to the environment that they have finally come to inhabit. The various states they assume will be reflected in the characteristic wavelengths of the radiation they emit or absorb, which are used by astronomers to seek them out. Detecting these baryons in their different guises is one of the key activities in observational astrophysics, and assessing their abundances is a cornerstone of cosmology.

We shall begin our inventory in our own neighborhood: the local universe. As will be seen later, the liaison between protons and neutrons, themselves formed just instants after the Big Bang led, via a process called primordial or Big Bang nucleosynthesis,[2] to an almost uniform mix of hydrogen and helium, with a small amount of deuterium, and traces of lithium and

---

[1] The term 'thermal history' refers to the evolution of the energy/matter content of the universe, from the Big Bang to the present time. This allows us to define the principal eras of the dominance of one constituent over another, or epochs when changes of state occurred, such as the recombination or re-ionization of the universe (see Appendices).

[2] Primordial or Big Bang nucleosynthesis (BBN) lasted from about 3 minutes to 20 minutes after the Big Bang. By this time, the temperature and density of the universe had fallen below that needed for nuclear fusion, preventing elements heavier than beryllium from forming, but at the same time enabling light elements such as deuterium to exist. Big Bang nucleosynthesis leads to abundances of about 75 percent hydrogen, about 25 percent helium-4, about 0.01 percent of deuterium, traces (about 10^{-10}) of lithium and beryllium, and no other heavy elements. It is usual to quote these percentages by mass, so that 25 percent helium-4 means that helium-4 atoms account for 25 percent of the mass; however, only about 8 percent of the atoms would be helium-4 atoms.

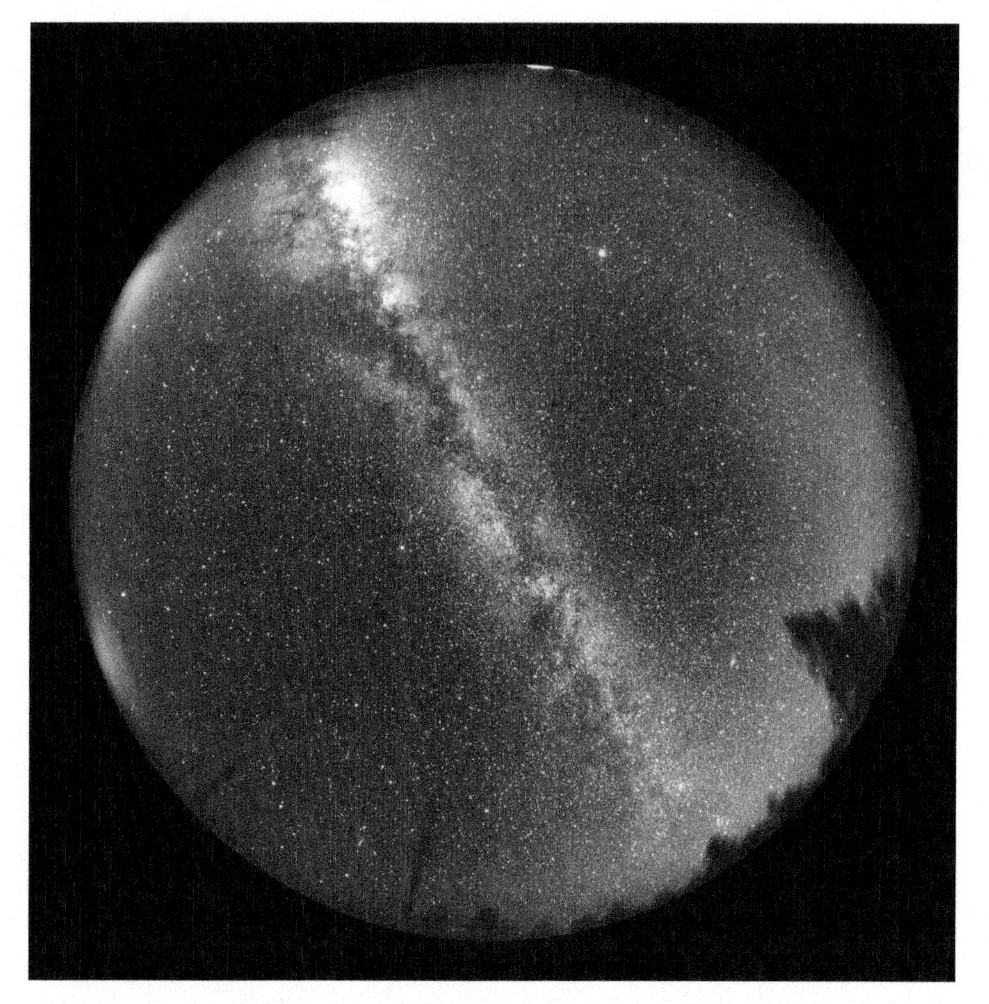

Figure 1.1 The summer Milky Way dominates the picture in this 20-minute exposure taken under the very dark skies atop the 10,720-foot summit of Mt. Graham in Arizona. The yellow glows on the horizon to the west are the lights of Phoenix and Tucson (Doug Officer and Paul Welch, Steward Observatory, University of Arizona).

beryllium. This uniformity was however not without its 'wrinkles', and it was unable to resist the action of gravity, which in time accentuated these minor heterogeneities. Now, much later, part of the baryonic matter has become concentrated into stars. Observing them is one way to 'weigh' the matter that formed them.

Counting the stars

Even though we now know that the Milky Way is in reality just one of hundreds of billions of galaxies, contemplating it on a clear summer's night is still a wondrous experience (Figure 1.1). Since our Solar System resides in the suburbs of our home Galaxy, we observe the Milky Way edge-on, projected upon the vault of the sky. So, almost at a glance, the eye of the beholder on that summer's night takes in more than 200 billion stars (although one will never be able to see more than about 2,500 of these with the naked eye). Who has not sometime looked up at the stars of night and wondered if we might ever be able to count them all? This little challenge, seemingly as impossible as counting the grains of sand on a beach, has nevertheless been taken up by astronomers. In fact, we can see many individual stars that are in our vicinity in the Galaxy, but there are two problems.

The first is that we cannot count stars one by one, with our eyes glued to the telescope. The Greek astronomer, geographer and mathematician Hipparchus, c.190 BC–c.120 BC, may well have recorded the positions and apparent brightnesses of about 850 individual naked-eye stars in his time (and in the unpolluted night sky of ancient Greece), but nowadays we prefer to secure images of our Galaxy with the aid of telescopes, and apply automated methods to achieve the count. However, the area covered by the Milky Way in the sky is very large, and obtaining an image of it in its entirety is therefore difficult. The second problem is that, even with the use of telescopes, there will always be extensive areas of the sky that are obscured, as if by dust on the lens. In reality, this dust is within the Galaxy itself, distributed in clouds, and consequently adding its contribution to the amount of matter while hiding what is behind it: this complicates the tally! Astronomers have found ways around these obstacles, and in the final analysis, by using photographic plates and their successors, CCD-camera images, it is now possible, with the help of computers, to count the stars.

An essential milestone in this census was reached in the early 1990s, thanks to the Hipparcos[3] satellite, launched on 8 August 1989 by the European Space Agency. The 'simple' mission of Hipparcos (which ran from August 1989 until February 1993) was to extend its ancient Greek namesake's work to encompass several hundred thousand stars, measuring their positions, distances, luminosities, and annual proper motions with great accuracy. The distances of the stars could be determined by the traditional astronomical method of parallax based on the motion of the observer. An

---

[3] Translator's note: Hipparchus/Hipparcos: Hipparchus is the usual English spelling for the Greek astronomer (although Hipparchos is more correct), but the satellite named after him is internationally called Hipparcos, which stands for High Precision Parallax Collecting Satellite..

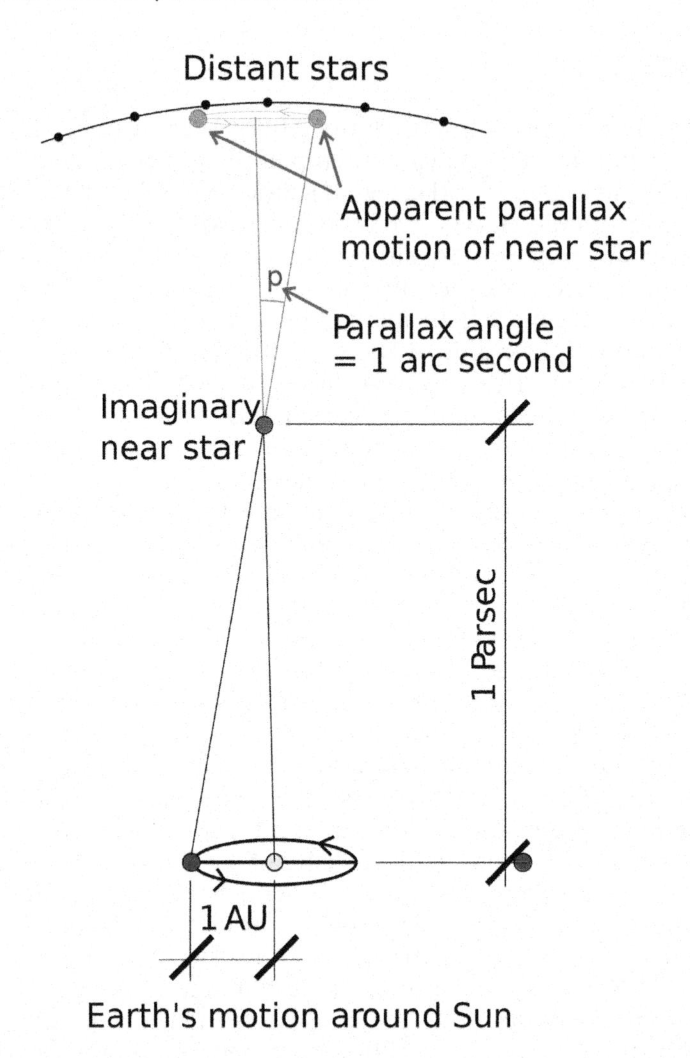

Figure 1.2 A nearby star, observed over six months as the Earth orbits the Sun, seems to have shifted its position in the sky against the apparently motionless distant background stars. The star has described an ellipse whose semi-major axis is its annual parallax. Knowing the distance from the Earth to the Sun, we can infer the distance to the star from its parallax. An imaginary near star having a distance of 1 parsec (3.26 light years) would have a parallax angle of 1 arcsecond. The genuinely nearest star system to our Sun, that of Alpha Centauri, has a parallax of 0.742 arcseconds.

outstretched finger at arm's length seems to jump from one side to the other compared with the background scenery, as we look at it first with the right eye and then with the left. In the same way, the nearest stars are 'seen' to change their position relative to more distant stars as the satellite observes them from the opposite ends of Earth's orbit around the Sun (Figure 1.2).

Fact box **More about the parallax method**

The notion of parallax can be intuitively explored by observing one's outstretched index finger, at arm's length, with first one eye and then the other. It will be seen that the finger seems to change position compared with background objects. Parallax is the effect of the changed point of view of the observer (which could be an artificial satellite) on the observation of an object. This notion has an essential role in astronomy, where parallax is the angle subtended by a given distance:

- the radius of the Earth, in the case of Solar System objects ('diurnal parallax');
- the semi-major axis of the Earth's orbit, i.e. one Astronomical Unit (AU), in the case of objects outside the Solar System ('annual parallax').

The nearer the object in question, the greater the apparent shift in position due to the displacement of the observer. The longest available baseline for the measurement of astronomical distances is the diameter of the Earth's orbit (2 AU), some 300 million kilometers. The distance at which the angular separation of the Earth and the Sun equals one second of arc is called a parsec, a unit of distance used in astronomy. One parsec equals 3.26 light years.

Before the launch of Hipparcos, parallaxes were known for about 8,000 stars only. Some years after the end of its mission, the high precision Hipparcos Catalog was published in 1997 containing parallaxes for almost 120,000 stars, to an accuracy of the order of one millisecond of arc; distances could be determined out to about 1,600 light years. The distances of 20,000 stars were determined to better than 10 percent and for 50,000 stars to better than 20 percent. A lower precision Tycho Catalog containing data for more than a million stars was published at the same time, while the enhanced Tycho 2 Catalog of over 2.5 million stars was published in 2000.

The European Space Agency's Gaia mission, due for launch in March 2013, will conduct a census of a thousand million stars in our Galaxy. This amounts to about 1 percent of the galactic stellar population. Monitoring each of its target stars about 70 times over a five-year period, Gaia will precisely chart their positions, distances, movements, and changes in brightness. Relying on the proven techniques of the Hipparcos mission, Gaia will repeatedly measure the positions of all objects down to magnitude 20 (about 400,000 times fainter than can be seen with the naked eye). Onboard object detection will ensure that variable stars, supernovae, other transient celestial events and minor planets will all be detected and cataloged to this faint limit. For all objects brighter than magnitude 15 (4,000 times fainter than the naked eye limit), Gaia will measure their positions to an accuracy of 24 microarcseconds. This is comparable to measuring the diameter of a human hair at a distance of 1,000 km. It will allow the nearest stars to have their distances measured to the extraordinary precision of 0.001 percent. Even stars near the galactic centre, almost 30,000 light years away, will have their distances measured to within an accuracy of 20 percent.

Using simple trigonometrical formulae, we can use the apparent changes in position to determine the distances to these nearby stars.

Thanks to measurements by the Hipparcos satellite, we now have the most complete picture yet of the structure and dynamics of our Galaxy. Enormous though this achievement may seem in its extent (see Fact box on previous page), accurate data have been obtained for less than a millionth of all the stars in the Galaxy, out to distances less than one fiftieth of the diameter of the Milky Way itself. It is true that a limited sample of people questioned for an opinion poll may not be representative of the population as a whole; and similarly, nothing proves that the region of our Galaxy we have sampled is representative of the whole. Given this problem, the question arises: how can we explore further?

The Milky Way: from mythology to science

On clear summer evenings in the northern hemisphere, the great pearly band arching across the sky is one of its most impressive structures. Its early mythological interpretation (as milk, spurting from the breast of the goddess Hera as she pushed away the infant Heracles) may no longer be current, but the name 'Milky Way' has survived since the time of the ancient Greeks. The generic term 'galaxy' is of similar origin, from the Greek word Ãáëáîßáò *Galaxias* (meaning 'milky').

It was the Greek philosopher Democritus (460 BC–370 BC) who first claimed that the Milky Way consisted of distant stars, but only when the first optical instruments became available was this apparently milky swath of light resolved into the myriads of stars of which it is composed. It is very difficult to determine the structure of the Milky Way, because we see it only from within. The first attempt to describe the shape of the Milky Way and the position of the Sun within it was made by William Herschel (Figure 1.3) in 1785. By carefully counting the number of stars in different parts of the sky, Herschel concluded correctly that the stars in the Milky Way were grouped into a huge disk-like formation, but his diagram showing the shape of the Galaxy (Figure 1.4) incorrectly depicted the Sun lying close to the center. Herschel also assumed (incorrectly) that all stars have the same intrinsic brightness and so only their distances would affect their apparent magnitude. Later, the discovery of different types of stars, with their different luminosities, caused astronomers to revise their methods. Accuracy was still not achieved, however, since the relationship between the temperature of a star and its luminosity is not an exact one. Only the parallax method could be used for accurate determination of a star's distance.

The first population of stars to be studied comprised those which are more or less the same as the Sun, having all been formed from the same gaseous environment, and all of similar chemical composition in spite of their varied masses. It soon became obvious that these stars were arranged in a flattened

Figure 1.3 Portrait of Sir Frederick William Herschel (15 November 1738–25 August 1822), the German-born British astronomer, technical expert, and composer. (Royal Astronomical Society, London.)

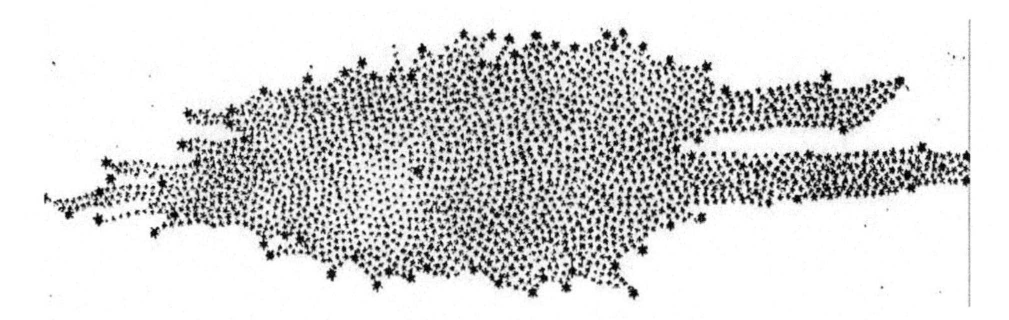

Figure 1.4 William Herschel's map of the Milky Way, based on star counts (from Wikipedia Commons, originally published in the *Philosophical Transactions of the Royal Society*, 1785).

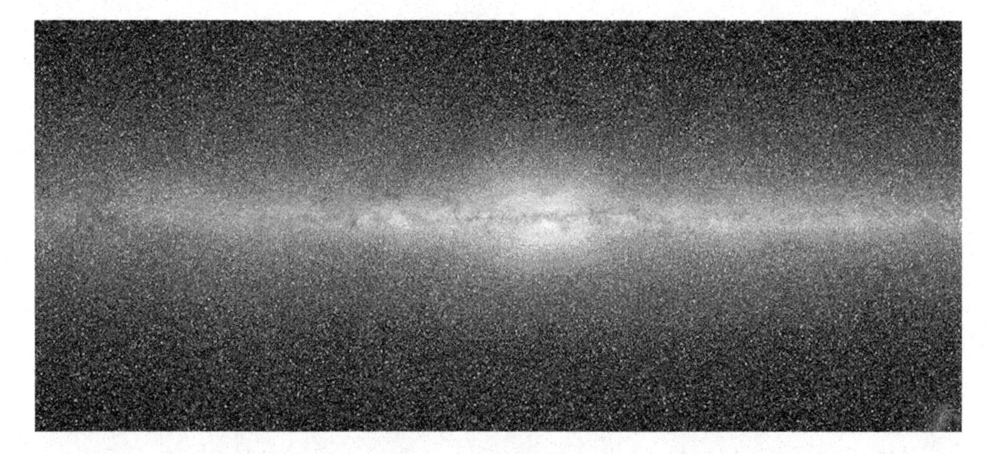

Figure 1.5 A near-infrared view of our Galaxy. This panorama encompasses the entire sky as seen by the Two Micron All-Sky Survey (2MASS). The image is centered on the core of the Milky Way, toward the constellation of Sagittarius, and reveals that our Galaxy has a very small bulge (normal for a late-type Sbc galaxy), which has the shape of a box or peanut (University of Massachusetts, IR Processing & Analysis Center, Caltech, NASA).

disk. In the direction at right angles to the milky band of the Galaxy, the vast majority of these stars were found to be less than 1,000 light years away, whilst in the direction of the Milky Way or in the opposite direction (away from the galactic center), distances could be as great as 20,000 and 10,000 light years, respectively. So the stars, it seemed, were arranged in a flat, disk-shaped structure (Figure 1.5). This disk is not uniform, and the spatial distribution of the youngest stars, like that of the HI gas clouds (detected at radio frequencies), revealed fluctuations in stellar densities, forming spiral regions around the center of the Galaxy. From all this, we may deduce that our Galaxy is a spiral system of some 200 billion stars about 100,000 light years in diameter, and that the Sun lies almost 27,000 light years from its center. Note that the mean density of matter is almost constant throughout the disk. Density waves, as they propagate, are responsible for triggering the formation of stars, causing the appearance and brightening of the spiral arms.

The exact nature of this spiral structure is still the subject of some debate among astrophysicists, since the detection of spiral structures seen edge-on is extremely difficult, and relies on detailed simulations of the distribution of matter. Observations are further complicated by the presence of dust clouds, which veil the stellar background by absorbing light. This problem may be resolved through the use of infrared observations, since infrared is not absorbed by dust. A first hurdle was overcome during the 1990s, thanks to ground-based observations that revealed the existence of a bar across the center of our Galaxy, transforming its status from 'spiral' to 'barred spiral'. Data from the Spitzer Space Telescope have recently much increased our knowledge of the spiral structure of the Galaxy. While previous models had suggested the existence of four spiral arms, it now seems that there are in fact only two main arms, surrounded by a few secondary structures (Figure 1.6).

In the course of his studies of different stellar populations during the first half of the twentieth century, the astronomer Walter Baade classified them into two major families according to the width of the lines present in their spectra. It later transpired that this distinction was due to the chemical abundance of 'metals' (elements heavier than helium). Metals in Population I stars - found in the disks of spiral galaxies, and particularly in the spiral arms - were comparable in their abundances with those in the Sun, and had therefore come from an environment already enriched with elements from previous generations of stars. However, Population II stars had come from a much less rich environment; they were of lower mass and much older (between 11 and 13 billion years old, as opposed to about 5 billion years for our Sun). Moreover, studies of the distribution of these stars showed that they are not scattered throughout the disk, but lie in a spherical halo that is difficult to detect because it displays no activity due to ongoing star formation. This is why our unaided eyes see only the disk.

The halo is thought to be the first structure formed, about 12 billion years ago, from almost primordial gas: hence its dearth of metals. Only the least massive stars remain of this first stellar generation; the others have already consumed their fuel and their lives have ended. The most massive stars, exploding as supernovae, have ejected gas enriched with heavier elements such as carbon, oxygen, and iron into the interstellar medium. The disk formed later from this material - enriched by the first generation of stars - from which new stars like the Sun, and the planets accompanying them, were formed. Nowadays astronomers have identified that the Milky Way is composed of several distinct parts that formed at different times and thus with different populations of stars, comprising the bulge, thin disk, thick disk, inner halo, and outer halo (Figure 1.7).

Figure 1.6 An artist's impression of how our Milky Way galaxy would appear as seen from above. It is possible to study the structure of the Milky Way both in the radio and far infrared parts of the electromagnetic spectrum whose wavelengths are not absorbed by the dust that makes it difficult to observe in the visible part of the spectrum. The infrared observations have been made by the Spitzer Space Telescope It is thought that there are two main spiral arms starting from opposite ends of a central bar. Our Sun lies in what is called the Orion Spur which lies between the two main arms below the centre of the image (NASA/JPL-Caltech).

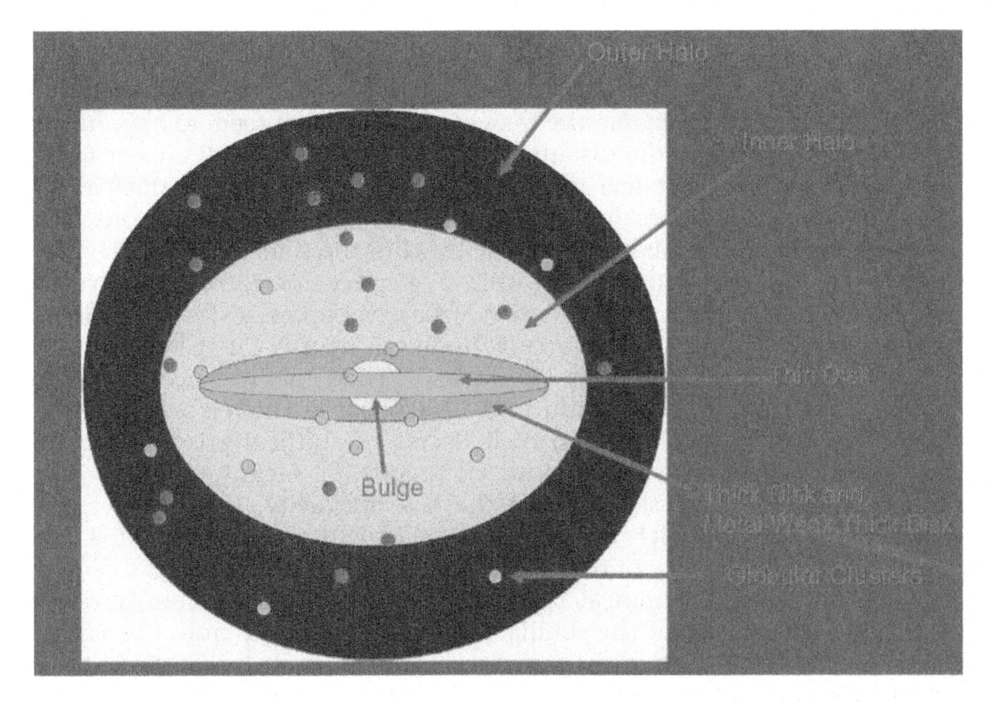

Figure 1.7 The Milky Way is composed of several distinct parts that formed at different times and thus with different populations of stars. **The Bulge** comprises the central spherical region of our Galaxy, which contains mostly old population II stars. In general, the stars have a moderately low metallicity compared with the Sun. It also is thought to harbor a massive black hole. **The Disk** is the flat disk-shaped part of the Galaxy, which contains the main spiral arms, and our Sun. It is dominated by Population I stars and is generally broken into two parts: the more metal-rich thin disk and the more metal-poor thick disk. **The Halo** is a spherical distribution of old, Population II stars that encompasses all the other parts of the galaxy. Recent work has suggested that the halo may be made of two distinct parts – *Inner Halo*: The older more metal-rich part of the halo that has little net rotation. It is thought to have formed during the first few large mergers that formed the Milky Way. *Outer Halo*: The younger and more metal-poor part of the halo that appears to have a net retrograde motion. It is thought to have been formed by later accretions of dwarf galaxies onto the Milky Way (Timothy Beers, University of Michigan).

Continuing the tradition of their 'back-to-front' nomenclature, astronomers call the very first population of extremely massive and hot stars to be formed from primordial gas 'Population III' or metal-free stars. These contain only those elements created during primordial Big Bang nucleosynthesis immediately following the Big Bang (see footnote 1 earlier in this chapter).

Farther afield in the Galaxy

Just as, when viewed from afar, the trees of a forest do not seem to have much dimension in depth, so the distant stars of the Milky Way all appear to be scattered at the same distance around the sky. When we observe this part of the stellar population, we lose any notion of volume. Now, in order to estimate the amount of matter contained within the stars of the Galaxy, we have to be able to count them within a given volume, i.e. we have to investigate their density. Mass density ρ may be expressed for example in solar masses per cubic parsec. Similarly, luminous density ℓ may be expressed in terms of solar luminosities per cubic parsec. Therefore, we need somehow to restore our vision in depth within the Galaxy, a feat which is achievable in the case of only a small fraction of its stars. The difficulty is resolved by astronomers resorting to calculations and modeling techniques. Based on sound principles of physics such as the law of gravity, they predict, for example, the way in which the number of stars within a small volume will depend upon the distance of the region in question from the galactic center. The validity of the theoretical representation (or model) is confirmed or disproved by the fact that the simulated galaxy, 'projected' onto the sky by the computer, must contain the same number of stars as the real Galaxy for any given region. A little mathematics will then lead to the final result of the exercise: the number of stars and the luminosity for the desired volume. A first estimate of the mass of all these stars can in fact be carried out by ascribing to them a mean value based on the nearest stars to the Sun. However, once we have arrived at this result, are we certain that we have indeed counted all the baryons?

In fact, to complete the picture, it will be necessary to include other galactic components that are also composed of baryons, i.e. also 'weighing' the dust, the gas, etc. These components are unfortunately not observable with instruments working only within the visible part of the electromagnetic spectrum. Nowadays, we observe our Galaxy at all wavelengths (Figure 1.8), and more specialized instruments are therefore required. For example, the COBE satellite, designed to measure cosmic background radiation, and the IRAS satellite, working in the infrared, have provided astrophysicists with detailed maps of the dust in the Milky Way (Figure 1.9).

Finally, and, as one can imagine, after much hard work, both observational and computational, astronomers have arrived at estimates of the quantity of baryonic matter in the Milky Way. Unfortunately, this kind of measurement is still far too localized, and is valid for our immediate environment, but not necessarily representative of the whole of the Galaxy, nor, on the larger scale, of the universe. Cosmologists cannot feel satisfied with this. As is the case with opinion polls as elections draw near, it is wise to seek out a more representative sample.

Figure 1.8 In recent decades, studies of our Milky Way have benefited from a tremendous broadening of coverage of the electromagnetic spectrum by ground-based and spaced-based instruments. Presented here are images of the sky near the galactic plane, derived from several space and ground-based surveys, in spectral lines and continuum bands spanning a frequency range of more than 14 orders of magnitude, from 408 MHz frequency radio waves (top) to gamma-rays (bottom). Each image represents a 360° false color view of the Milky Way within 10° of the galactic plane; the images are in galactic coordinates with the direction of the galactic center in the center of each image. Very dusty areas appear clearly on the optical (visible light) image, in the form of dark regions. Immediately above, appear the old stars, for which the near infrared emission is not absorbed by dust. The next two images above (infra-red and mid-infra-red, having a longer wavelength than near infra-red) show the emission of the dust, included in a very fine disk, less than 500 parsecs in thickness compared with its diameter of 30,000 parsecs. The structures above and below the disk are the clouds closest to the Solar System, which thus appear at a different angle. The images above, at still longer wavelengths, show the emission by molecular gas which is present in zones of star formation. On the images at the bottom (X-rays and gamma rays) appear the most energetic objects (pulsars, supernova remnants, etc.). It is seen that the absorption by dust towards the galactic center manages to absorb even X-rays, whereas the emission of the central black hole appears in gamma rays (in the center of the bottom image). For scale, the vertical dimension of each image is forty times the angular diameter of the full Moon on the sky; the areas shown represent about one-sixth of the entire sky (National Space Science Data Center at NASA Goddard Space Flight Center).

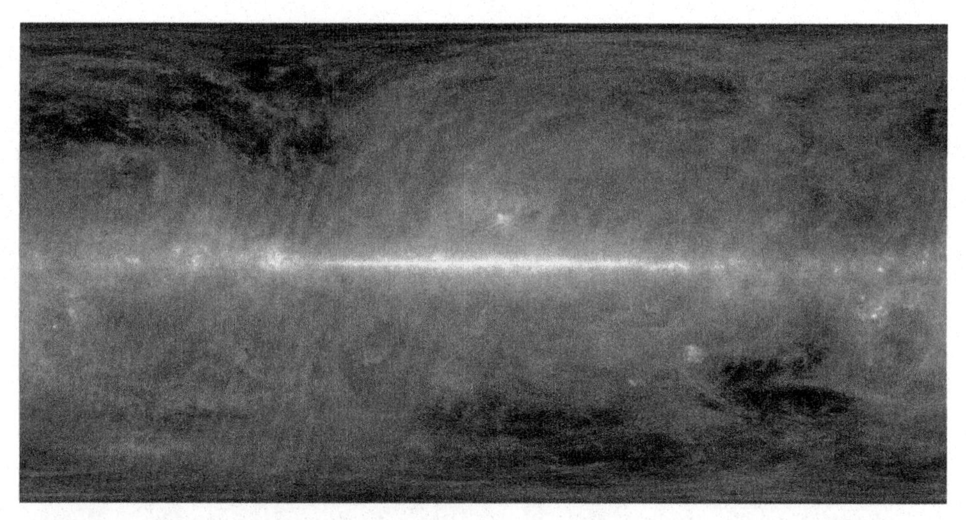

Figure 1.9 This image shows the dramatically different view of our sky when viewed in far infrared light. At this wavelength of 100 microns, the light from stars is imperceptible against the cool glow of vast dust clouds warmed by the stars' light. In visible light such dust blocks our view of more distant stars and galaxies, but in far-infrared light it glows dramatically. The dust in the Milky Way is mostly distributed in a disk that is very thin. This bright narrow band shows most of the dust in our Galaxy. The fainter filaments above and below this disk represent only the very closest dust clouds as seen from our vantage point in the middle of this disk. The image has been constructed from data collected by the Infrared Astronomical Satellite (IRAS) and the Cosmic Background Explorer (COBE). The two datasets have been merged and individual point sources have been removed to highlight the dust structures themselves (IRAS/COBE).

Meaningful relationships

So our first efforts at 'valuation' have not told us everything, but we have nevertheless made progress in our research. One thing we have realized is that mass and luminosity are the distinctive signs of a given stellar population. We can sum this up in a single number: the mass-luminosity relationship (MLR),[4] which is a kind of identity badge for a given population of stars. If we can identify the proportions of different stellar populations in a galaxy, from observations at various wavelengths, we will be able to translate our

---

[4] The mass-luminosity relationship (MLR) is obtained from the mass density ρ and the luminosity density ℓ, which are, by definition, mass M and luminosity L divided by volume V. The MLR can therefore be written: $\text{MLR} = \rho V/\ell V \equiv \rho/\ell$.

measurement of the global luminosity of a distant galaxy into its mass by simple multiplication. Job done!

Also, cosmologists like to normalize mass densities by introducing an essential quantity of cosmological models: the critical cosmological density. Written as ρ_c, this is the critical density separating the 'open' and 'closed' models of the universe. Its value is $\sim 1.6 \cdot 10^{11}$ solar masses per Mpc^3 or $\sim 10^{-29}$ gm per cm^3 for the cosmological concordance model.

Rather than use the idea of density itself, the parameter Ω_b is defined,[5] the indicator *b* signifying that baryons are involved. It is this normalized value Ω_b (i.e. baryon density divided by critical cosmological density) that will be compared with the other contributions of the matter/energy content of the universe. Finally, it is advisable to subdivide the contribution of the baryons according to the different forms they take (stars, gas, dust...), forms to which a particular Ω_b parameter can be ascribed. Also, these partial values can be compared to the total quantity of baryons in the cosmos, a value that will be evaluated in chapter 4.

So we return empty-handed from our first foray into the cosmos, because, obviously, we have stayed too close to our galactic home. Exploring a district does not give us a full knowledge of a whole town, and we will know even less about the land and its inhabitants. We really must get out more...

---

[5] Ω is a parameter used in cosmology, corresponding to the density of a component of the matter/energy content of the universe divided by the critical cosmological density ρ_c. The accompanying indicator shows the component in question (e.g. Ω_b corresponds to the contribution by baryons).

2 The realm of the nebulae

"In the Kingdom of Heaven all is in all, all is one, and all is ours."
Meister Eckhart

In order to 'weigh' the ordinary matter of the universe, it is clearly not enough just to estimate the contribution of the baryons through the study of a single astronomical object, however near it may be. This is our inevitable conclusion, but fortunately, the Milky Way has many siblings in the sky, making it possible to adopt a more global approach to the population as a whole.

Island universes

This term, together with 'nebulae', was formerly used to designate what we now refer to as galaxies. That these are systems exterior to our own Milky Way was finally recognized at the time of The Great Debate[1] between astronomers Harlow Shapley and Heber D. Curtis. Leaving aside those cataloged as irregulars (which were in the majority in the early universe), these systems have characteristic shapes, allowing them to be classified into two great families: spiral galaxies (both ordinary and barred) are disk-shaped, with a central bulge of stars, and elliptical galaxies look like soccer balls or rugby balls (Figure 2.1). How these cosmic denizens were formed, their properties, and especially the existence of the three main families, continue to be the subjects of intensive research. It is, however, established that the driving force playing an essential role in their gestation is that of gravity, causing dark matter and baryons to congregate until a galaxy is born; and it is gravity that still maintains their stability and creates the morphologies already mentioned.

As this process of formation proceeded, the disk-shaped galaxies came to acquire rotational motion: a boon for astronomers, since, using Newton's Laws of gravitation, they can deduce the total mass of a spiral system from its

---

[1] The Great Debate revolved around the question of whether the observed 'nebulae' were extragalactic or not. See http://antwrp.gsfc.nasa.gov/diamond_jubilee/debate.html

Figure 2.1 Some examples of different types of spiral and elliptical galaxies. (Upper left) The Sc type spiral galaxy NGC 5457 (M101) in Ursa Major. (Lower left) The SBb type barred spiral galaxy NGC 1300 in Eridanus. It is thought that our Milky Way is a SBc type barred spiral galaxy. (Upper right) The more or less spherical E0 type elliptical galaxy NGC 4458 in Virgo. (Lower right) The more ellipsoidal E5 type elliptical galaxy NGC 4660 in Virgo (NASA/STScI and the Hubble Heritage Team (STScI/AURA)).

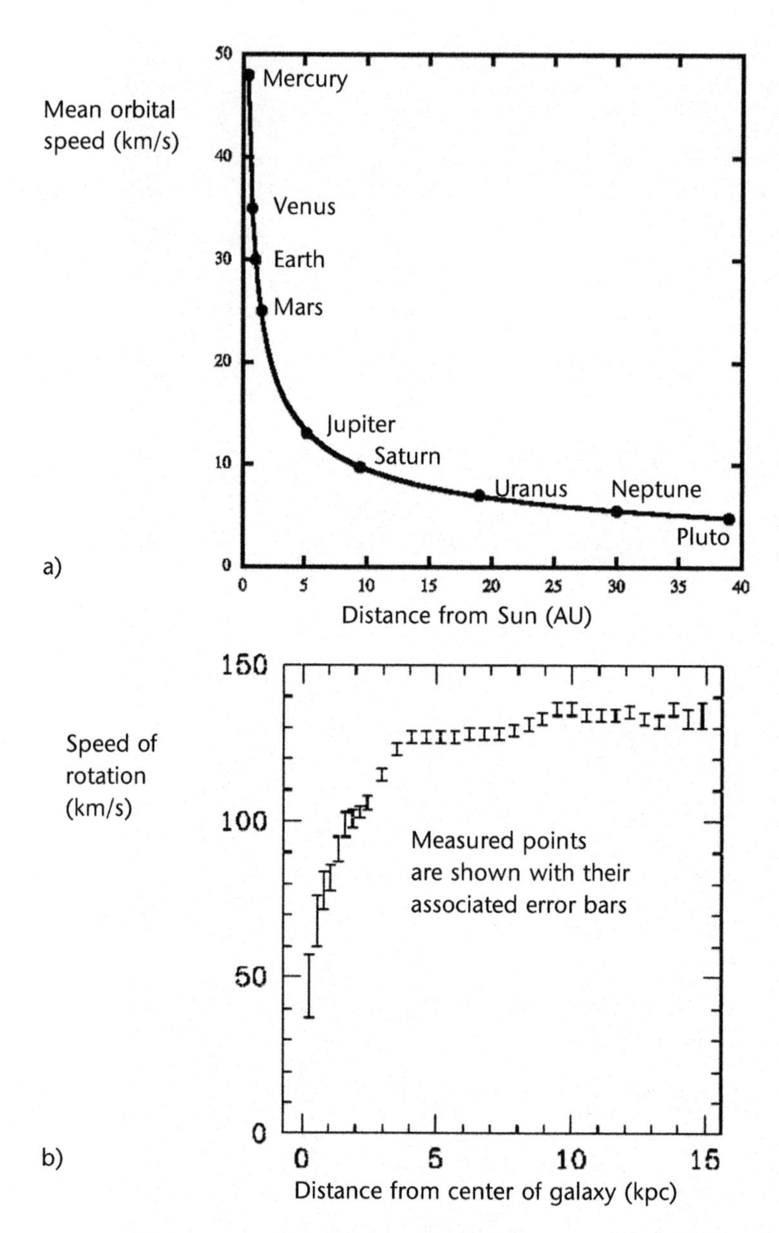

Figure 2.2a) Orbital velocities of the planets around the Sun as a function of their solar distances. The curve matches that predicted by Newton's theory of gravitation, if all the mass is situated within the planetary orbits. **b)** Rotation curve of a spiral galaxy, i.e. variation in rotational velocity with distance from the centre of the galaxy. Velocities are determined from the shifts in wavelength (Doppler Effect) in the radiation emitted by different zones of the galaxy. The obvious difference between this curve and that of the planetary orbits shows that the mass is not concentrated in the central parts, where most of the stars are to be found.

observed rotation. By careful measurement of the rotation curves[2] (Figure 2.2b) of these objects and the application of gravitational laws, their masses can readily be determined. The mass arrived at by this method is known as the 'dynamical mass', to distinguish it from estimates determined by other methods. This is an essential value, since it is a reliable indicator of the total mass of galaxies, obtained without having to deal with the contents of these objects in detail. So we can 'weigh' spiral galaxies, and also the other types, without having to count all their component stars.

As we shall see later, comparison of the results obtained by the different techniques will bring about consequences both unexpected and fundamental...

Iceberg galaxies

We encounter a problem when observing external galaxies: they are ever more distant and ever fainter, and therefore increasingly difficult to observe. What is more, because of these great distances, we can no longer see or count their individual stars. There is, however, one advantage - their apparently small size in the sky means that they can easily be imaged in a single exposure, and we can immediately estimate their total luminosity. Moreover, if these objects lie near enough to us in the universe, we can easily study their properties in various regions of the cosmos. This is the strategy (the establishment of a database of 100 million external galaxies) successfully adopted by researchers pursuing the Sloan Digital Sky Survey (SDSS). The SDSS program is one of the most ambitious and influential surveys ever realized in the history of observational astrophysics.[3] The SDSS uses a dedicated 2.5-meter telescope at Apache Point Observatory, New Mexico (Figure 2.3), equipped with two powerful special-purpose instruments. The 120-megapixel camera images 1.5 square degrees of sky at a time, about eight

---

[2] Spiral galaxies rotate, and the velocity of any region of a galaxy can be measured by studying the apparent change in frequency (known as the Doppler Effect) of the light emitted. Measurement of the change in frequency (or wavelength) gives an indication of the rotational velocity. This velocity can be compared to the distance from the center of the galaxy to establish a rotation curve. Generally, velocity reaches a plateau at a considerable distance from the center (Figure 2.2b). Spiral galaxies do not spin as if they were solid bodies (like, for example, a spinning top), with velocities proportional to distance from the center), and neither do they mimic the motion of the planets around the Sun, where velocity decreases with distance (Figure 2.2a).

[3] For further information, see http://www.sdss.org.

Figure 2.3 Image of the 2.5-meter Sloan Digital Sky Survey telescope at Apache Point Observatory in New Mexico (from the website of the SDSS http://www.sdss.org/gallery/).

times the area of the full Moon. A pair of spectrographs fed by optical fibers measure spectra of (and hence distances to) more than 600 galaxies and quasars in a single observation. A custom-designed set of software pipelines keep pace with the enormous data flow from the telescope.

Over eight years of observations by the telescope dedicated to the program (SDSS-I, 2000-2005; SDSS-II, 2005-2008), with 40 researchers ensuring coordinated data management and analysis, SDSS obtained deep, multi-color images covering more than a quarter of the sky and created three-dimensional maps containing more than 930,000 galaxies and more than 120,000 quasars. Meanwhile, SDSS is continuing with the Third Sloan Digital Sky Survey (SDSS-III), a program of four new surveys using SDSS facilities. SDSS-III began observations in July 2008 and will continue operating and releasing data through 2014.

The SDSS team's precise spectroscopic studies have enabled them to determine simultaneously the positions, luminosities and stellar-population details of many thousands of galaxies, and thereby the typical MLRs, thought to be universal. This hypothesis is certainly a valid one, given the area covered by the SDSS, which ensures a truly representative sample. With a great deal of telescope time and a great volume of perspiration behind them, the researchers have used the phenomenal mass of data acquired to estimate the amount of baryons concentrated within the stars ($\Omega_{b\text{-}stars}$, adapting the notation Ω_b already encountered in the previous chapter).

It will be seen immediately from Table 2.1 and Figure 2.4, when comparing this estimate of the baryon content of galaxies with the total contents as expressed by Ω_{tot}=100 percent, that the contribution of the stars to the whole is a feeble one: about 0.25 percent of the whole, and a tiny fraction of the ordinary matter of the universe. But this is by no means the end of the story. Above, we mentioned how the total mass of galaxies can be measured by, for example, analyzing the rotation curves in the case of spirals. This allows us to determine what we have called the dynamical mass, which can also be measured for a sufficiently representative sample of these objects. The broad

Table 2.1 Fractions of ordinary (or baryonic) matter existing as stars and gas within galaxies, compared with total energy/matter in the cosmos, expressed as different parameters of Ω_b.

Nature of ordinary matter	Contribution to total energy/matter content of universe	
Stars	$\Omega_{b\text{-stars}}$	~0.25 percent
Gas in atomic form (hydrogen, helium)	$\Omega_{b\text{-HI + HeI}}$	~0.06 percent
Gas in molecular form	$\Omega_{b\text{-HII}}$	~0.016 percent
Total		**~0.32 percent**

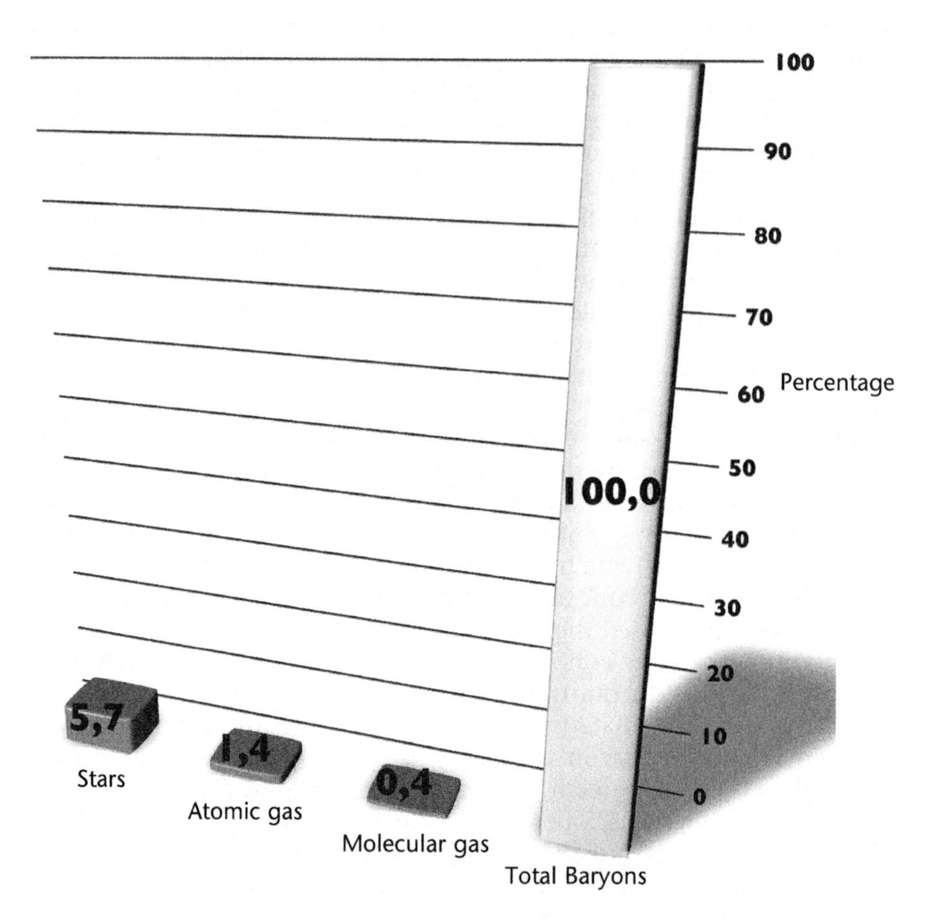

Figure 2.4 Comparison of quantities of ordinary matter as detected in the stars, atomic gas and molecular gas within galaxies, and the total contents of this baryonic matter in the cosmos. It is readily seen that a very large fraction is unaccounted for.

comparison of this estimate of the total mass of galaxies with that of the baryons detected shows a deficit in the latter of a factor 5-10. What can be the reason for such a discrepancy? Either there are many baryons hidden within galaxies which have evaded detection in spite of the scrupulous surveys of the astrophysicists, or some other explanation will have to be found... So we will have to widen our investigations if we wish to sign off our cosmic accounts.

Galactic fauna

So where are the missing baryons hiding? As in the case of the Milky Way, we must also bring into the reckoning the ordinary matter present in the form of gas and dust, or in any other form that does not emit visible light: gas, both atomic and molecular, and also dust (a term used to describe small grains of matter in the interstellar medium, usually present as silicates, the scientific word for sand). The gas, essentially the atomic hydrogen present in the interstellar medium, is very dilute, and cold. It emits no visible light (unlike, for example, the gas in a neon tube, which is dense and excited by the difference in electric potential to which it is subjected). The only way in which this interstellar gas can be detected is through absorption (the gas absorbs light from a background star at characteristic wavelengths, and the composition and density of the gas can then be determined), or through emission, *via* the so-called 'fine-structure' transition of atomic hydrogen. This transition gives rise to radiation at a wavelength of 21 centimeters (radio waves), so radio telescopes must be used to detect this gaseous component. A similar problem is found in the case of molecular gas, whose transitions are less energetic than those of its atomic counterpart. Here too, radio telescopes are indispensable. Various surveys have revealed a whole range of cosmic 'fauna' within the interstellar medium: from simple molecules like carbon monoxide, water and methane, to very complex organic types containing, for example, twenty carbon atoms, sometime arranged in rings, such as benzene.

All these detailed studies have enabled astrophysicists finally to identify the various contributions of these elements. It seems a little paradoxical to summarize the results of all this considerable expenditure of time and energy in just a few numbers in a table: such is the scientist's lot... But the most frustrating thing about the data so painstakingly acquired is that the contributions of the different elements are, in the final analysis, so small: just a few per cent (Table 2.1 and Figure 2.4), and the value we arrive at for the density parameter of the baryons present in galaxies, in the classic forms of stars or gas, is only about 0.32 percent!

In conclusion, in spite of a seemingly complete inventory, the discrepancies mentioned have not been accounted for. On the one hand, baryons in galaxies do not constitute the majority of the energy/matter content of the cosmos, and on the other, they do not even seem to constitute the mass responsible for the observed rotation curves of spiral galaxies (not to mention other kinds). This gulf between the measured total and (in the broad sense) luminous masses of the galaxies (and the same is true of clusters of galaxies) leads us to infer the existence of hidden mass or dark matter,[4] in order to be able to explain the discrepancy. Seen in this way, luminous matter is only the visible 'tip of the iceberg' of matter as yet undetected directly. The frustrating

---

[4] Hidden mass or dark matter is probably matter composed of particles as yet hypothetical, but envisaged by certain extensions of the model of particle physics. The idea of dark matter is the response to the problem of hidden mass, a problem brought about by the analysis of the dynamics of galaxies and clusters of galaxies and the existence of gravitational 'mirages'.

conclusion has in fact engendered new questions. What is the nature of this dark matter? Might there be, for example, baryonic matter that has so far eluded us?

Stars and MACHOs

Observations of spiral galaxies, and indeed of clusters of galaxies, show indisputably that their dynamics, i.e. the rotation of the spirals and the motions of the galaxies within their groups or clusters, cannot be explained in terms of their visible contents alone. So, we can either call into question the laws of gravitation, or infer an abundant amount (about 80 percent of the total mass) of hidden matter. The first hypothesis has not been favored by physicists, although it has not been abandoned. If the second is to be pursued, then what kind of matter can it be that keeps itself hidden from astronomers?

The stars so far mentioned are, by definition, luminous bodies. Their light originates in thermonuclear reactions deep within them. We know for example that our Sun, which has been shining for some five billion years, will continue to shine for a further five billion years, since the fusion of hydrogen in its core can last that long. However, for these nuclear reactions to be triggered, conditions of density and temperature at the center of the star must be fulfilled. It has been shown[5] that a star's mass is the most important factor that determines its lifetime. Paradoxically, the more massive a star the shorter its lifetime.

For a star to be stable, the weight of its successive layers must be counterbalanced by high pressure from within, preventing it from collapsing inwards. If this pressure is high enough, the nuclei of atoms which under normal conditions in a gas will repel each other because of electrostatic repulsive forces, may be fused together, emitting energy which is then radiated away by stars. Knowing the relationship between pressure and temperature, we can deduce that a star needs to be of a mass at least equal to one-hundredth of the mass of the Sun for this cosmic reactor to be switched on. If this mass is not attained, the nuclear fires will not be ignited, and the body in question will not have energy to radiate. Such dark bodies are sometimes known as 'failed stars', but a better label for them is brown dwarfs (Figure 2.5) or MACHOs.[6] It is certainly tempting to consider MACHOs as

---

[5] See, for example, *Exploding Superstars* by A. Mazure and S. Basa (Springer/Praxis, 2009).

[6] MACHO is the abbreviation for MAssive Compact Halo Object, as opposed to the WIMP (Weakly Interacting Massive Particle), also a candidate for dark matter.

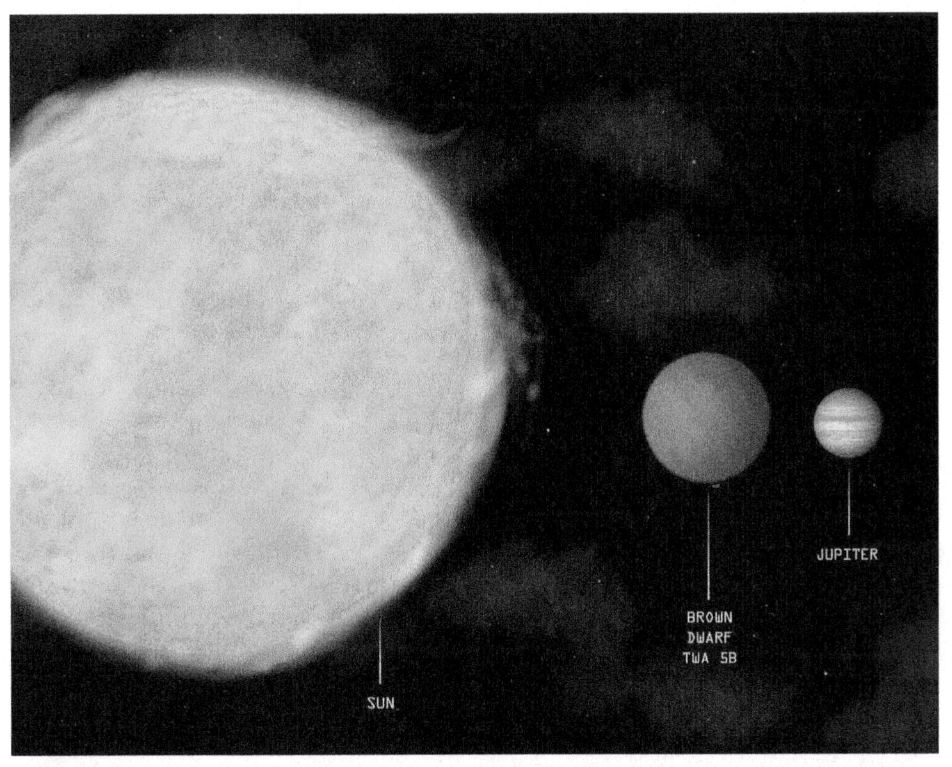

Figure 2.5 The approximate size of a brown dwarf (center) compared with the Sun (left) and the giant planet Jupiter (right). Although brown dwarfs are similar in size to Jupiter, they are much more dense and emit infrared radiation, whereas Jupiter shines with reflected light from the Sun (NASA/CXC/M. Weiss).

potential candidates for the coveted title of 'missing mass', or 'dark matter'. They are, by definition, dark stars (at least as far as visible radiations are concerned, since their temperatures, typically a few hundred degrees, mean that they emit only in the infrared). They are probably very common objects, since it seems that, in general, there are greater numbers of stars among the less massive types. So what they lack in size, they could make up for in number, and their total mass might be enough to constitute the hidden mass in the form of dark baryons.[7]

---

[7] Since the hunt for baryons has proved in part fruitless, the hypothesis of dark baryons has arisen: these emit no radiation and are therefore practically undetectable. They may exist in the form of brown dwarfs or as very cold molecular hydrogen.

The saga of the dark baryons

In the 1980s, a Polish researcher called Bohdan Paczynski achieved fame by demonstrating that the brown dwarfs scattered throughout the galactic halo might be involved in gravitational (micro-)lensing effects. We know that, according to Einstein and his theory of General Relativity, the motions of celestial bodies can be explained not by gravitational forces between them but rather by the local curvature of space caused by the presence of mass at a given location. Now, this applies not just to the paths of the planets, but also to electromagnetic radiation and to the particles that constitute it: photons. So, a beam of light will be deflected in the vicinity of a massive object, as if it were being viewed through an optical lens. The two phenomena are closely analogous, whence the term 'gravitational lensing'. These lensing effects can be highly spectacular when photons emitted by a more distant galaxy encounter a cluster of galaxies composed of several hundred members, with a total mass of as much as a hundred thousand billion (10^{14}) times that of the Sun. The distorted image of the galaxy is seen as if through the bottom of a glass bottle. The relative positions of the remote object(s), the lensing cluster of galaxies, and the observer may combine to create magnificent gravitational arcs or lensed arcs (Figure 2.6). These distorted and amplified images of the

Figure 2.6 This Hubble Space Telescope image of a rich cluster of galaxies called Abell 2218 is a spectacular example of gravitational lensing. This cluster of galaxies is so massive and compact that light rays passing through it are deflected by its enormous gravitational field. This phenomenon magnifies, brightens, and distorts images of more distant objects (Andrew Fruchter (STScI) et al., WFPC2, HST, NASA).

distant galactic light sources (sometimes called mirages) are caused by the presence of the intervening cluster of galaxies along the paths of the photons emitted. In accordance with the laws of gravity, the rays of light will be curved by massive objects. Not only are the images of the distant objects distorted, but they are also amplified, appearing brighter due to the concentration of the light rays (Figure 2.7). Therefore, the intervening clusters of galaxies are sometimes known as 'gravitational telescopes'.

The phenomenon of lensing may occur in the case of objects other than distant galaxies and intervening clusters. For example, a brown dwarf in the halo of our Milky Way galaxy may act as a micro-lens upon the light from a star within the Large Magellanic Cloud, one of the Milky Way's satellite galaxies. This would seem a worthwhile way of detecting potential MACHOs. However, since these objects are in constant motion within our Galaxy, the phenomenon of amplification will be but a transient event to an observer on Earth. Fortunately, there are millions of stars within the Large Magellanic Cloud which might be momentarily 'brightened' by Galactic brown dwarfs. Encouraged by the potentialities of this technique, several large-scale surveys (e.g. EROS, AGAPE, MACHOs, OGLE, DUO...) were begun during the 1990s, especially in the direction of the Large Magellanic Cloud. One difficulty however is that, in spite of the enormous numbers of objects in play (several million), the probability of observing the effect remains fairly low. The expected rate is only a handful of detections per million stars, and the duration of the phenomenon depends on the mass of the lensing body and the relative motions of both 'lens' and source. The phenomenon can last from some tens of days in the case of a lens of the order of a tenth of a solar mass, to one day for a lens of less than a millionth of a solar mass.

Surveying the Large Magellanic Cloud, the Andromeda Galaxy or the central regions of our own Galaxy, as several groups have done, requires large imaging systems (Schmidt cameras or CCD arrays). Also, powerful computers are needed to compare the thousands of images, hunting for some fleeting variation in the brightness of a star. The analysis is further complicated by the existence in the cosmos of all kinds of objects whose brightness varies. This difficulty is countered by the simultaneous use of a range of filters when observing, since the phenomenon of gravitational lensing is independent of wavelength, unlike a great number of other transitory stellar phenomena.

It would seem therefore, after several years of work studying tens of millions of stars, that the mass represented by brown dwarfs contributes but little to the total mass of ordinary baryons. Dark baryons are therefore not the 'white knights' coming to the rescue of hidden matter, and the question already posed remains: are there other kinds of missing baryons?

Figure 2.7 Astronomers have used the technique of gravitational lensing to discover one of the youngest galaxies in the distant universe. The Hubble Space Telescope was first to spot the newfound galaxy. Detailed observations from the W.M. Keck Observatory on Mauna Kea in Hawaii revealed the observed light dates to when the universe was only 950 million years old. The distant galaxy's image is being magnified by the gravity of a massive cluster of galaxies (Abell 383) parked in front of it, making it appear 11 times brighter, one of the consequences of gravitational lensing (NASA, ESA, J. Richard (Center for Astronomical Research/ Observatory of Lyon, France), and J.-P. Kneib (Astrophysical Laboratory of Marseille, France), with thanks to M. Postman (STScI)).

More about galaxy clusters

Clusters of galaxies are the largest and most massive stable gravitationally-bound structures to have arisen so far in the process of cosmic structure formation in the universe (Figure 2.8). They typically contain from a few hundred to 1,000 galaxies (if there are fewer than 50, they are known as groups of galaxies), together with extremely hot X-ray emitting gas and large amounts of dark matter. Their total masses may be of the order of 10^{14} to 10^{15} times that of the Sun, contained within a volume typically 5 million to 30 million light years across. Their considerable mass makes them privileged targets for studying the nature and distribution of matter in the cosmos. Within such a cluster, the spread of velocities for the individual galaxies is about 800–1000 kilometers per second.

Clusters of galaxies were first identified in the course of galaxy counts using Schmidt cameras. Astronomers George Abell and Fritz Zwicky came to realize that, in some localized areas of the sky, there were significantly greater

Figure 2.8 Hubble Space Telescope image of the magnificent Coma Cluster of galaxies, one of the densest known galaxy collections in the universe. Hubble's Advanced Camera for Surveys viewed a large portion of the cluster, spanning several million light years across. (NASA, ESA, and the Hubble Heritage Team (STScI/AURA).)

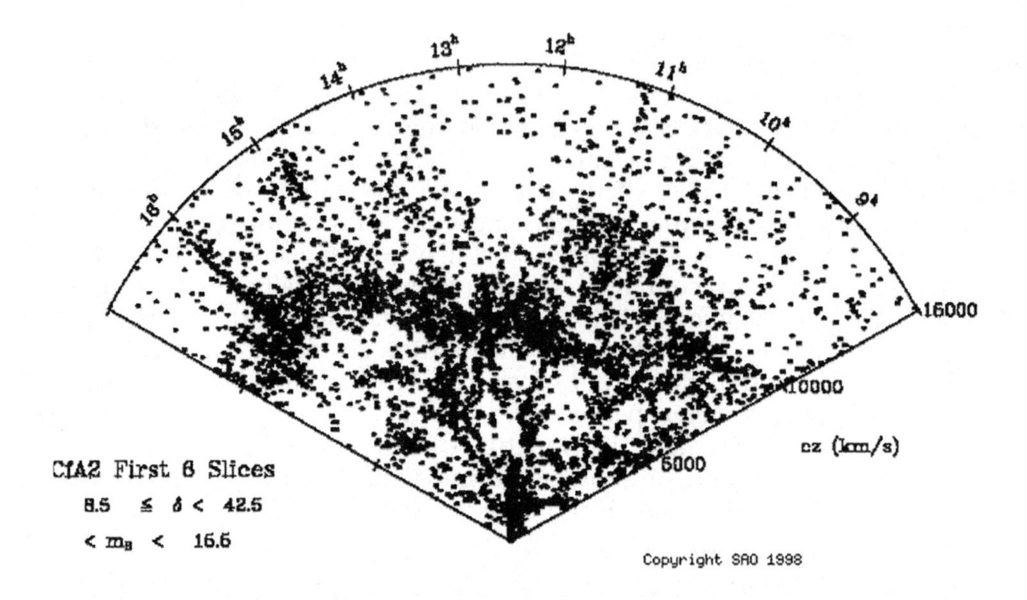

Figure 2.9 In the 1980s, the Harvard Smithsonian Center for Astrophysics initiated one of the first large surveys of the universe with the aim of mapping the distribution of galaxies by measuring their velocities v, and thereby their distances d, using Hubble's Law ($v = Hd$). This revealed the existence of vast voids, filaments and 'walls'. We also see large elongated structures, all in the direction of the observer: these correspond to clusters of galaxies containing several hundred objects, and we are measuring the radial component of their velocities within the cluster.

numbers of galaxies than were recorded elsewhere. Thus, they showed the existence of clusters and groups of galaxies, nowadays listed in two great catalogs which bear their names.

The galaxies within the cluster are in equilibrium due to gravitational effects essentially created by dark matter. They all pursue their trajectories through the cluster at velocities reaching several hundred kilometers per second. Astronomers can measure these motions with the aid of spectrographs, using the Doppler Effect caused by the apparent shift in the wavelength of the radiation emitted by each galaxy.

This proper motion complicates the representation of clusters: the distances of the galaxies are in fact derived from their radial velocities (in the direction of the observer) using the formula for the expansion of the universe. However, if a fraction of this velocity represents velocity within the cluster, then the measured distance is falsified. This is why clusters appear elongated on charts such as that shown in Figure 2.9. To measure the internal velocity dispersion of a cluster, it is necessary to establish the distance common to all the galaxies (the size of the cluster being small compared with

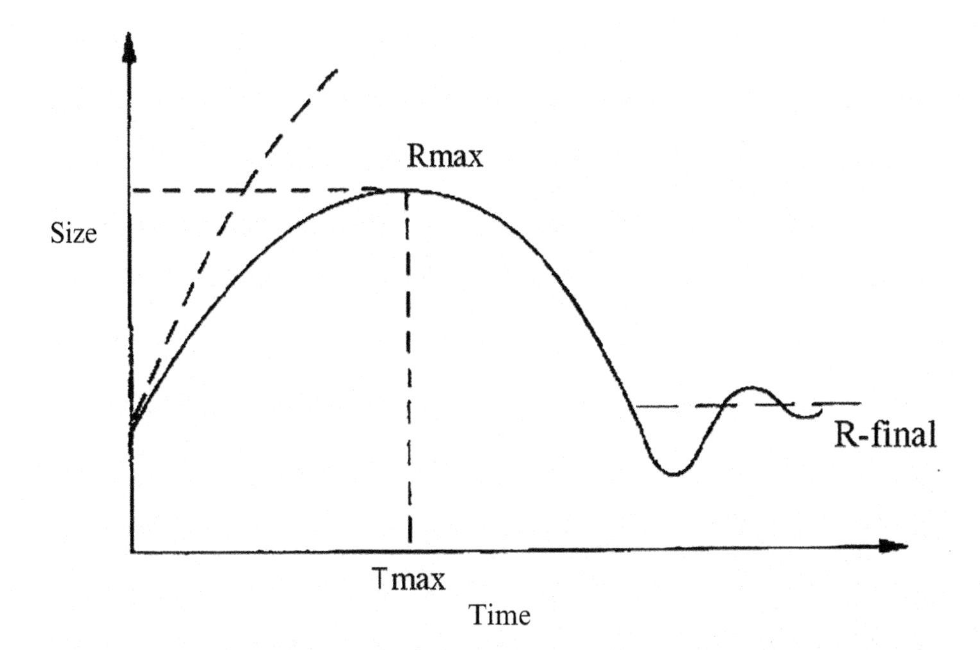

Figure 2.10 When a structure forms, it must first of all decouple from the expansion. Here we see the radius around a fixed mass M, which will become a distinct structure. Time elapses towards the right. At first, the mass continues to expand (solid line), at a rate which slows in relation to the mean rate of the expansion of the universe (dashed line). The relative density becomes greater and greater relative to the rest of the universe. The time will come (T_{max}) when the density reaches the critical density that allows the structure to collapse upon itself. Then the expansion will be reversed until the structure reaches virial equilibrium, and kinetic energy is balanced by potential energy. During the collapse, the velocities of agitation are higher, until the equivalent 'pressure' balances out the forces of gravity. We then obtain a secular stable structure, of radius R_{final}.

its distance from us), then deduce the internal velocity from the measured velocity and the overall velocity of the cluster.

The impressive development of multi-object spectroscopy during the 1990s meant that the velocities of several hundred or even of thousands of objects could be measured simultaneously, and the universe could thus be surveyed directly and not just through stellar images. These new-style surveys allowed clusters of galaxies to be identified and cataloged.

Clusters of galaxies, like galaxies themselves, originated in the growth of 'clumps' initially present in the cosmic fluid. Under the effects of gravitation, a small excess in density increases with time (equivalent to redshift), such that at a given moment it becomes large enough for the region of space concerned to 'separate' from the general expansion, at redshift z_{sep} (Figure 2.10). The forming system will continue, of course, to follow the expansion

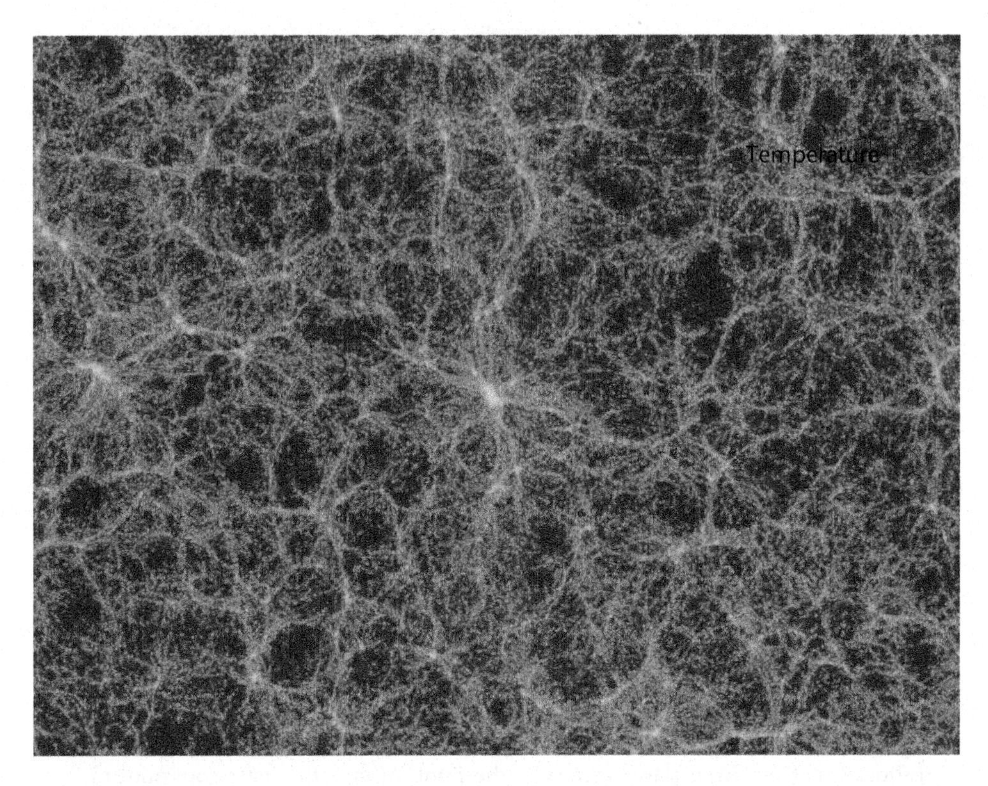

Figure 2.11 Large-scale structure of the universe. A slice of a computer simulation of our universe by the Virgo Consortium. The nodes, filaments and walls of light make up the large scale structure that we can observe in outer space.

generally, but will for some time experience internal oscillations before reaching a definite stability at size R_V. The system has become stable, or what is known as virialized, with virial radius R_V of the order of a few million light years.

Clusters and their contents (dark matter, galaxies and hot gas) are globally in equilibrium, held together by the common gravitational potential essentially created by dark matter. In the large-scale distribution of structures, clusters of galaxies appear as 'knots' at the intersection of filaments, and which themselves surround large empty voids (Figure 2.11). In fact, the peripheral zones of clusters are continuously fed with galaxies and gas 'flowing' along these filaments, gradually populating the cluster to the accompaniment of violent effects (tidal effects, interactions with the gas between clusters...) within the environment that they encounter. Only at the cores of clusters is there any real stability.

We now know that, in spite of their name, clusters of galaxies are composed mostly of dark matter ($\sim$ 80 percent), X-ray emitting gas ($\sim$ 15 percent) at temperatures of 10 million to 100 million degrees, and finally

Figure 2.12 Composite image of a collision between two clusters of galaxies. Superposition of X-ray emissions from hot gas and the projected mass in the galaxy cluster 1E 0657-56, familiarly known as the 'Bullet Cluster'. (X-ray: NASA/CXC/CfA/M.Markevitch *et al.*; Lensing Map: NASA/STScI, ESO WFI, Magellan/U.Arizona/D.Clowe *et al.*; Optical: NASA/STScI, Magellan/U.Arizona/D.Clowe *et al.*) The mass distribution projected on the sky corresponds to the mass of the clusters as reconstructed by gravitational lenses (deformations of background galaxies). Two clusters can easily be distinguished. The smaller cluster, at right, seems to have traversed, in the manner of a cannon ball, the larger one at left. During this collision, the hot gas of the subcluster has encountered the hot gas of the large cluster, and has been braked such that the two gaseous areas are closer together than the two masses. To the right we clearly see a shock wave in the shape of an arc, resulting from the passage of the smaller cluster. The projected X-ray gas and masses are also superimposed on the optical image showing the individual galaxies. The different behaviour during the collision of the hot gas and the stellar masses, including dark matter, allows us to separate the three components and test the models (after Clowe *et al.*, 2006).

galaxies ($\sim$ 5 percent). With the development of multi-wavelength observational techniques, it has become possible to analyze and identify in detail the contents of these systems, and even to 'visualize' their diverse components. For example, the object which has become known as the 'Bullet Cluster' turns out to be the result of a collision between two clusters. This system has been observed in detail at both visible and X-ray wavelengths (Figure 2.12).

Figure 2.13 These two NASA Hubble Space Telescope images reveal the distribution of dark matter in the center of the giant galaxy cluster Abell 1689, containing about 1,000 galaxies and trillions of stars. The galaxy cluster resides 2.2 billion light years from Earth. Hubble cannot see the dark matter directly. Astronomers inferred its location by analyzing the effect of gravitational lensing, where light from galaxies behind Abell 1689 is distorted by intervening matter within the cluster. Researchers used the observed positions of 135 lensed images of 42 background galaxies to calculate the location and amount of dark matter in the

cluster. They superimposed a map of these inferred dark matter concentrations, tinted blue (above), on an image of the cluster taken by Hubble's Advanced Camera for Surveys (facing page). If the cluster's gravity came only from the visible galaxies, the lensing distortions would be much weaker. The map reveals that the densest concentration of dark matter is in the cluster's core (NASA, ESA, D. Coe (NASA Jet Propulsion Laboratory/California Institute of Technology, and Space Telescope Science Institute), N. Benitez (Institute of Astrophysics of Andalusia, Spain), T. Broadhurst (University of the Basque Country, Spain), and H. Ford (Johns Hopkins University)).

Here we see (in red) the distribution of the hot gas, revealed through its X-ray emissions as detected by NASA's Chandra X-ray Observatory. This X-ray gas represents, essentially, the baryons within the cluster. The galaxies and the stars they contain, as seen on the image, in fact constitute only a small part of the total baryonic mass, and this itself is only a small fraction of the total. Most of this mass is dark matter (blue), which is of course undetectable directly since it emits no radiation. It manifests itself through gravitational lensing effects upon background galaxies, effects which allow us to reconstitute and visualize its distribution within the cluster. The first striking thing about this image is the superposition of the distribution of galaxies upon that of the dark matter, both spatially shifted with respect to the distribution of the X-ray emitting gas. This results from the fact that these diverse components do not react in the same way to the collision between the two clusters. This is a great boon for astrophysicists, meaning that they can 'see' separately, and 'weigh', the different contributions to the mass. Also, the analysis of this cluster is a very strong argument in favor of the existence of dark matter, since theories of modified gravity (MOND) cannot easily account for the phenomenon.

Astronomers using the Hubble Space Telescope have also derived the distribution of dark matter in the center of the giant galaxy cluster Abell 1689. This cluster contains about 1,000 galaxies and trillions of stars and lies 2.2 billion light years from Earth. Again, astronomers inferred the location of the dark matter by analyzing the gravitational lensing effects, as light from galaxies behind Abell 1689 is distorted by intervening matter within the cluster. Researchers used the observed positions of 135 lensed images of 42 background galaxies to calculate the location and amount of dark matter in the cluster. If the cluster's gravity came only from the visible galaxies, the lensing distortions would be much weaker. The map reveals that the densest concentration of dark matter is in the cluster's core (Figure 2.13).

As well as galaxies and dark matter, clusters contain hot gas, which is essentially left over from the initial gas involved in its formation. Like the galaxies, the gas is in equilibrium within the common gravitational potential, which means that it acquires sufficient energy to reach temperatures of the order of many millions of degrees, becoming ionized and emitting X-rays corresponding to this range of temperatures. With instruments of the sensitivity of those carried by XMM-Newton or Chandra, it is possible to carry out large-scale surveys of the sky in which groups and clusters of galaxies appear as zones of extended emissions, easily distinguishing them from stars, galaxies or quasars which can also be sources of this range of wavelengths.

Photons of 'fossil' cosmological radiation at 3 degrees Kelvin (the cosmic background radiation or CMB) bathe the cosmos, including its clusters of galaxies. They therefore interact with the hot intragalactic plasma, and in particular with the electrons of this plasma. These photon-electron interactions tions cause the energy and therefore the frequency of the photons of the

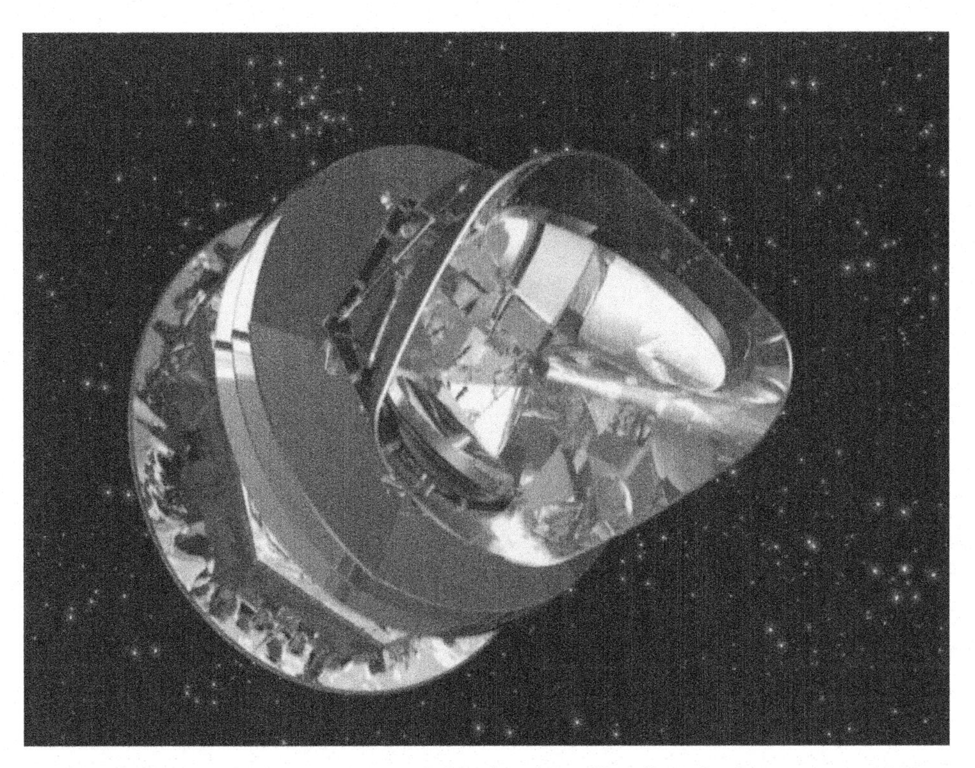

Figure 2.14 Artist's impression of the Planck satellite, launched on 14 May 2009 (together with the Herschel space telescope), on a mission to map the anisotropies of the cosmic microwave background radiation field over the whole sky, with unprecedented sensitivity and angular resolution. Planck was formerly called COBRAS/SAMBA. After the mission was selected and approved (in late 1996), it was renamed in honor of the German scientist Max Planck (1858-1947), Nobel Prizewinner for Physics in 1918. Fifty days after launch, Planck entered its final orbit around the second Lagrangian point of the Sun-Earth system (L2), at a distance of 1.5 million kilometers from Earth.

CMB to be modified. This so-called 'Sunyaev-Zel'dovich effect' can be measured at millimetric wavelengths, and the Planck satellite (Figure 2.14), launched in 2009, has detected, during its first year of observations, about 200 clusters of galaxies in this way.

Finally, whatever method of detection is used, we can count clusters in space, and at various spectral redshifts. These counts, as a function of volume or redshift, constitute a powerful test of cosmological models and the determination of associated parameters.

3 Getting warmer...

Some like it hot

It may seem regrettable, but progress in the technologies used by scientists, and even some of their discoveries, are often associated with political or military ventures. Such was the case with spaceflight, initiated during the Second World War (although rockets were in fact being used by the Chinese as weapons as early as 1200) and reaching a climax during the Cold War. Another, and striking, example concerns the discovery of bodies of unknown type, labeled gamma-ray bursts. This was an unexpected result of spying on the USSR by American satellites during the 1960s.[1]

At that time, the two great powers had signed a treaty banning nuclear tests in the atmosphere. However, the USA feared that such tests might be held in space, or even on the far side of the Moon. Its mistrust led it to attempt to verify any possibility of clandestine tests in the USSR or in space through the detection of the resultant gamma-rays, the most energetic domain of the electromagnetic spectrum (Figure 3.1). Detectors on the Vela series of satellites did indeed record such events, but they turned out on analysis to be coming from cosmic objects, not from the ground or from nearby space. A new kind of astronomy was born.

The epic of space

From the time of Galileo and his early telescope, progress in astronomy was, until the twentieth century, essentially ensured by the increasing size of telescopes, set up at mountainous sites with ever clearer views, and by improvements in optical components and photographic emulsions. This kind of astronomy, however, was always the astronomy of 'visible' radiations. The invention of radar during the Second World War and the later conversion of military detectors to more peaceful uses opened a new window upon the universe, ushering in the era of radio astronomy. There followed a fundamental discovery, that of the cosmic microwave background radiation, by Arno Penzias and Robert Wilson in 1965. However, other parts of the

---

[1] See *Exploding Superstars* by A. Mazure and S. Basa (Springer/Praxis, 2009).

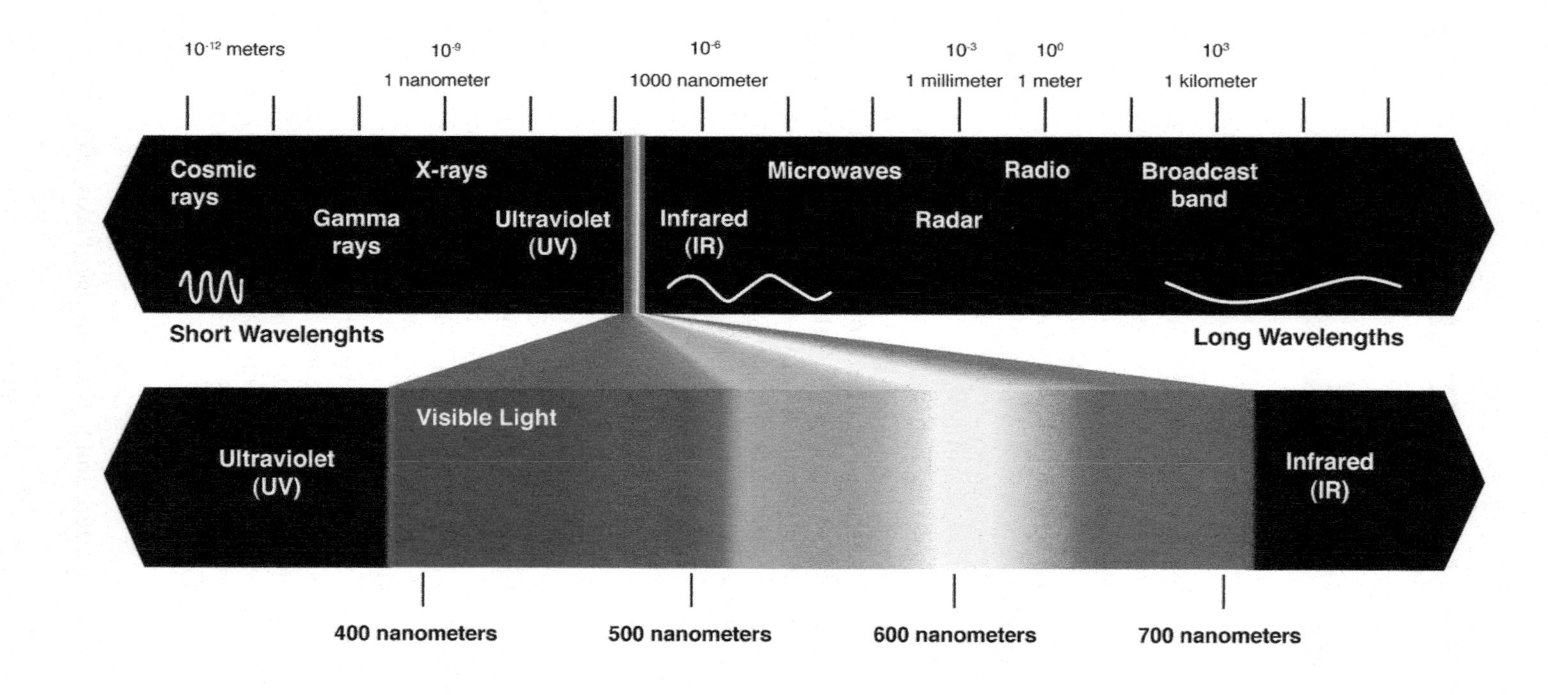

Figure 3.1 The electromagnetic (EM) spectrum extends from the lowest frequencies used for radio transmissions, through to higher frequency, shorter wavelength radio waves, microwaves, infrared radiation, visible light, ultraviolet radiation, and X-rays to gamma radiation at the ultra-short-wavelength end. The EM spectrum covers wavelengths from thousands of kilometers down to a fraction of the size of an atom. Hotter, more energetic objects and events create higher energy radiation than cool objects. Only extremely hot objects or particles moving at very high velocities can create high-energy radiation like X-rays and gamma-rays.

electromagnetic spectrum were not so easily observed. The atmosphere forms a barrier to, among others, X-rays and gamma-rays, protecting life on Earth, and the only way to 'see' through this barrier is to go into space. Astronomy in space began with the V2 rockets captured in Germany at the end of the Second World War, their warheads replaced by scientific instruments; the first objective was the study of the upper atmosphere and meteorological research. The first scientific rocket (the WAC Corporal) was launched in 1945, reaching an altitude of about 60 kilometers (Figure 3.2).

In 1948, researchers at the US Naval Research Laboratory were the first to detect X-rays from the Sun. As new generations of more powerful launch vehicles followed, similar programs were soon undertaken in Europe: in France, with the Véronique launcher, and in Britain with the Skylark. It was however the USSR that, in October 1957, sent the first artificial satellite, Sputnik 1, into space. The conquest of the cosmos was indeed under way.

X-Ray telescopes and detectors

X-rays, discovered in 1895 by the German physicist Wilhelm Roentgen, belong to the domain of wavelengths of the electromagnetic spectrum between 0.01 Angström and 10 Angströms (0.1-1 nanometers). These wavelengths correspond to energy levels of the order of 1eV to 1MeV, according to the relationship $E = h\nu$ for photon energy, where the frequency ν is given by $1/2\pi\lambda$.

These energies are sufficient to ionize matter and even split molecules (one application of this being the sterilization of food). Such ionizing radiations are therefore evidently dangerous for living creatures, but if properly controlled, they have shown themselves to be useful in radiography and radiotherapy. As well as these beneficial (or dangerous) applications, X-rays have joined the range of methods used by astronomers to explore the universe ever since the discovery that most celestial bodies emit in this spectral domain.

Even the Sun emits X-rays (Figure 3.3), but they pale into insignificance beside those emitted by the real 'stars' of this domain, bodies undergoing violent physical processes, for example active galaxies, black holes, supernovae, gamma-ray bursts and cataclysmic variable stars. Collapses and subsequent explosions, the ejection of jets of matter, accretion: all these processes are usually accompanied by the emission of X-rays.

X-ray photons are easily absorbed by matter, which explains their use in medicine for both observational and curative purposes. A few sheets of paper will suffice to block radiation in the range 0.5 to 5 keV. Therefore, X-rays do not penetrate the atmosphere, which is why X-ray astronomy was born at the same time as space flight was developed: finally, we could break through the barrier that at the same time both protected us and partially hid the universe from us. A new era had begun, at first through the use of balloons and rockets,

Figure 3.2 The first rockets. The WAC Corporal rocket in 1945. Next to it stands Frank Malina, a pioneer of space exploration and the first director of the Jet Propulsion Laboratory. Progress made in rocketry can be appreciated when we consider that a modern rocket like the Ariane 5 is more than 50 meters high and weighs 500 tons (10 percent of the weight of the Eiffel Tower) at launch (Jet Propulsion Laboratory, California Institute of Technology).

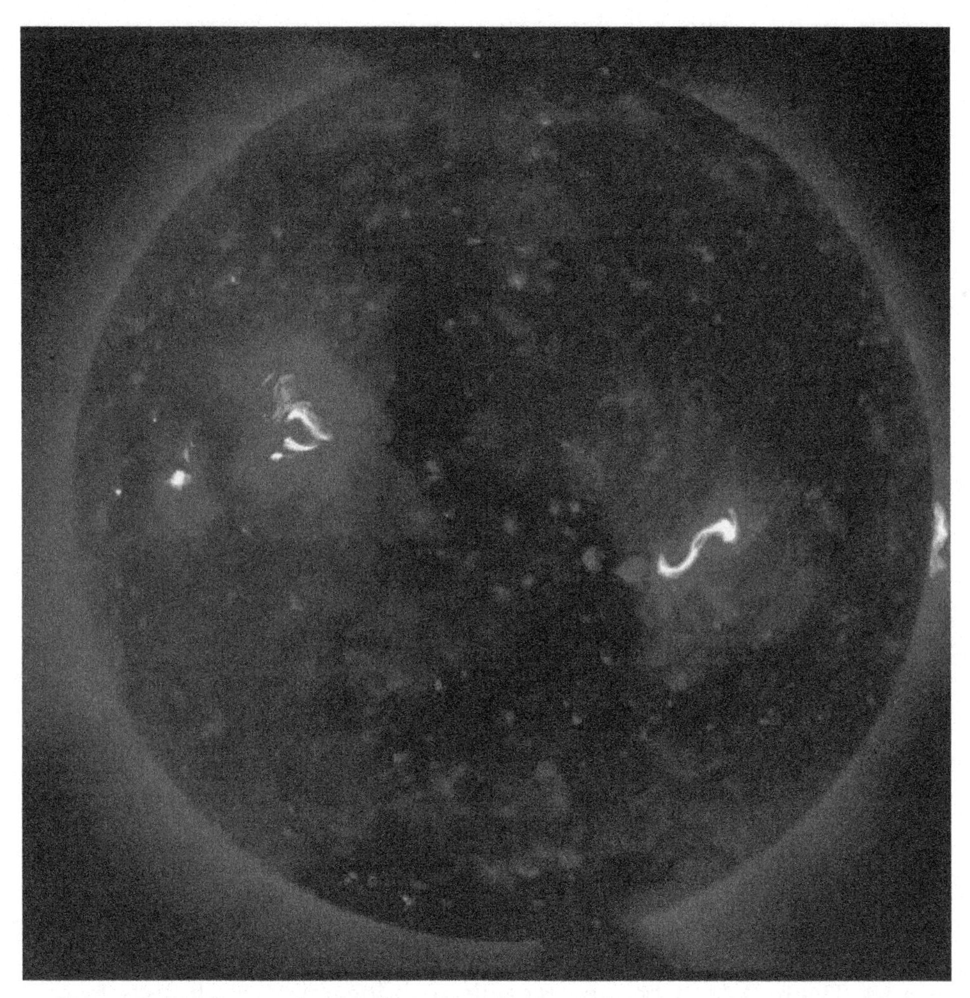

Figure 3.3 A feature that has been found to be common to many major solar eruptions is an "S" shaped structure in the solar active region immediately before the eruption takes place. Such an "S" shape – or sigmoid – can be seen in this full-Sun X-ray image, from the Hinode X-Ray Telescope (XRT). The bright sigmoid was observed on the right-hand side of the solar disk, just at the beginning of an eruption on 12 February 2007. Sigmoid structures may be observed for several days before the occurrence of an eruption.

and later with satellites. However, getting into space was not the only problem to be overcome.

Unlike their visible counterparts, which are reflected towards the detector more or less at right angles by the mirror of the telescope observing them, X-ray photons are totally absorbed as they strike the detector. So it was necessary to invent a new kind of optics, using new materials able to reflect

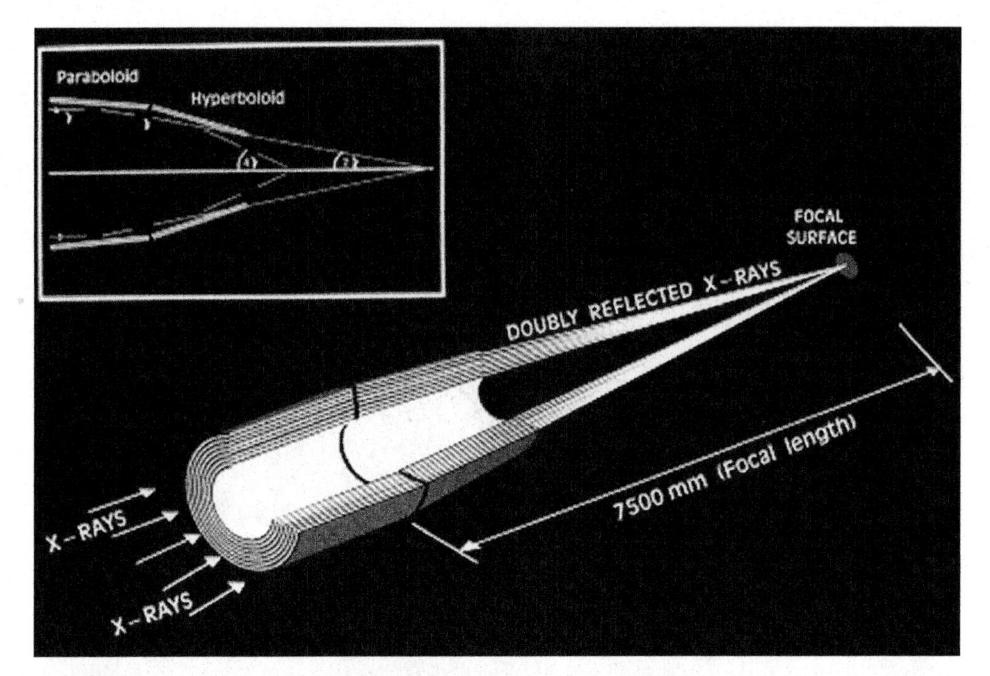

Figure 3.4 A diagram of the XMM-Newton X-ray telescope, showing the grazing incidence of the X-rays and the consecutive use of parabolic and hyperbolic mirrors to focus the beams. The overall weight of the satellite is 4 tonnes, and it is 11 meters tall.

rather than absorb this radiation, with a view to building specialized telescopes using a grazing incidence technique for cosmic X-ray photons. This led to an unexpected 'gold rush', in the sense that this metal possesses the required qualities. Using combinations of parabolic and hyperbolic mirrors (Figure 3.4), it became possible to bring together photons from a celestial source, allowing its position to be determined (although with much less accuracy than could be obtained in the optical domain). By counting photons on the appropriate detector (and taking account of the emitted flux and energy of the photons, arriving in effect one by one at the detector), an image of the source, and not just its luminosity, is obtained.

X-ray astronomy counts photons one by one (Figure 3.5), while in optical astronomy, innumerable photons are received simultaneously by the detector. Fortunately, these X-ray photons are very energetic, and are therefore able to interact quite easily with carefully selected detector material. However, they must not be hindered by the environment around this detector, which means that X-ray detection is no easy task. There are two main kinds of X-ray detector:

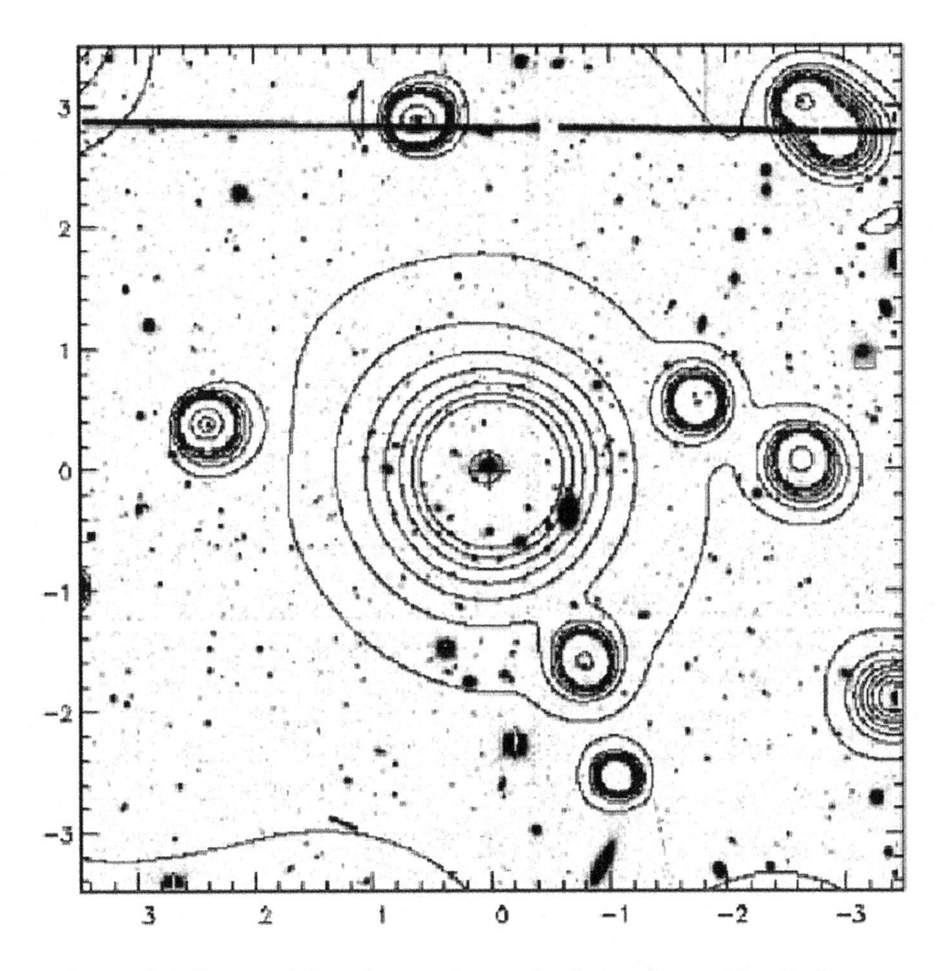

Figure 3.5 X-ray emissions from a cluster of galaxies observed by the European Space Agency's XMM-Newton satellite. The continuous curves represent the levels of intensity of this emission. This whole chart is based on no more than two hundred photons. The instrument collects approximately one X-ray photon emitted by the hot gas every minute. Fewer photons are received than there are galaxies in the whole cluster!

- *Scintillation* detectors measure light emitted when X-ray photons interact with atoms in a gas, causing the re-emission of visible photons. Essential technological advances have led to the development of charge-coupled devices (CCDs), formerly found only in the optical domain, capable of detecting emissions at these wavelengths.
- *Calorimetric* detectors measure the heat produced when X-rays are absorbed.

In fact, the kind of detector used depends upon the aim of the experiment: imaging requiring reasonable directional accuracy, or spectral recognition requiring very precise measurement of the energies involved.

Galaxies and 'hot' clusters

The first X-ray source detected outside our Galaxy was M87 (Figure 3.6), a galaxy in the Virgo Cluster, the nearest cluster of galaxies to our own. This was soon followed by the detection of similar sources in another neighboring galaxy cluster, in Perseus. So galaxies too were now recognized as X-ray sources, and many of them soon revealed themselves to be 'active' galaxies or quasars, within which black holes heat the matter around them by accretion[2] in their intense gravitational fields (Figure 3.7). The gravitational energy acquired by this matter as it is ingested by the massive black hole is radiated as X-rays. However, it soon became apparent that certain extragalactic emissions did not come from point sources such as stars or quasars (the zone of emission of a quasar is of relatively small size), but were in fact from extended areas of the sky. Early observations were carried out by instruments mounted on balloons and rockets, and the flights were of very short duration; the amount of data they returned was too limited for any firm conclusion to be drawn. The exact nature of these sources was revealed during the 1970s with the launching from Kenya of the first dedicated X-ray satellite, Uhuru (Swahili for 'Freedom').

It was then realized that these extended sources corresponded to clusters of galaxies already known from optical observations in the constellations of Virgo and Perseus. The emissions therefore came from regions of the universe of the order of several millions of light years across, and of luminosities equivalent to several billion times that of the Sun in the visible domain. These extended emissions became the object of intense scientific speculation, and X-ray astronomy began to come into its own with the advent of successive generations of satellites (the Ariel and HEAO series), culminating in 1978 with the X-ray observatory named after the father of relativity: Einstein. The Einstein Observatory's new capabilities, in particular its focusing X-ray telescope, led to a 'quantum leap' in what had been a discipline still finding its way.

It now became possible to show that all the clusters of galaxies (and, to a lesser extent, groups of galaxies) were emitting a considerable quantity of

---

[2] Accretion is the accumulation of matter as a result of interaction. Black holes, with their strong gravity, attract material from their immediate surroundings, forming an accretion disk.

Figure 3.6 Hubble Space Telescope Advanced Camera for Surveys image of the giant elliptical galaxy M87, the first X-ray source to be detected outside our Galaxy. This 120,000-light-year-diameter galaxy lies at a distance of 54 million light years. It is the dominant galaxy at the center of the Virgo cluster of galaxies, which contains some 2,000 galaxies altogether. M87 is the home of several trillion stars, a supermassive black hole, and a family of 15,000 globular star clusters (NASA, ESA, and the Hubble Heritage Team (STScI/AURA) with thanks to P. Cote (Herzberg Institute of Astrophysics) and E. Baltz (Stanford University)).

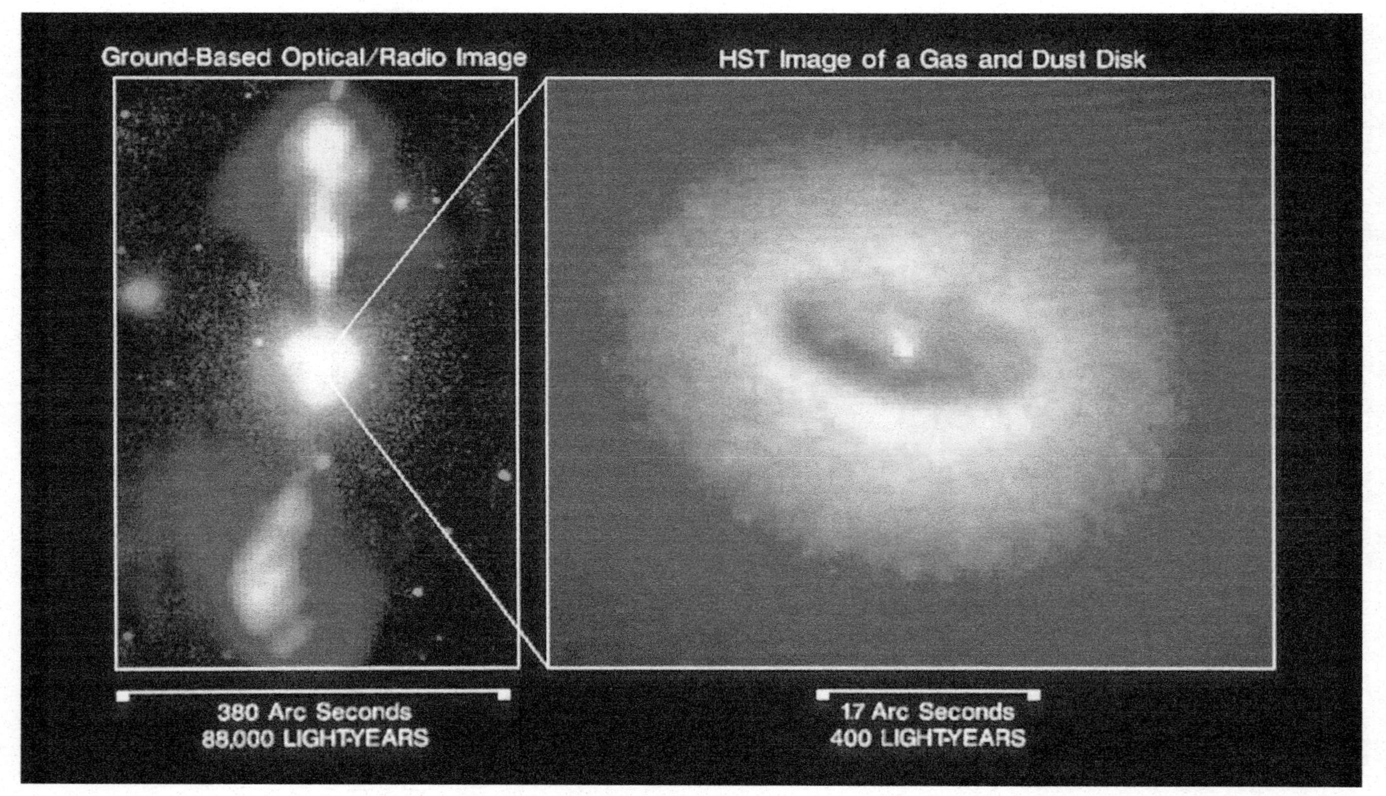

Figure 3.7 A black hole and its accretion disk. (Left) A composite optical and radio image of the elliptical galaxy NGC 4261, a member of the Virgo cluster of galaxies, 45 million light years distant. In the visible domain, this galaxy appears as a disk. The radio image reveals two jets issuing from the centre of the object. (Right) Hubble Space Telescope image of the core of this galaxy. It shows the presence of a gigantic disk of cool gas and dust, feeding into a central black hole. The disk has been accreted by the black hole, drawing in matter from its surroundings (NASA/STScI and ESA).

X-rays, across extended regions rather than from individual galaxies (Figure 3.8). The only explanation is that these clusters contain a considerable quantity of very hot gas, responsible for the observed emissions. More recently, European instruments have made their mark alongside American and Japanese counterparts (see chapter 8), carrying out detailed analysis of the nature of this plasma. It seems to be the remnant of the original vast cloud of gas from which the clusters and galaxies within them were formed. Dominated by the mass of the dark matter, its main constituent, the cloud has collapsed upon itself as a consequence of its own gravity. Individual galaxies were formed during this relatively chaotic process, until an overall equilibrium was established within the system. Tens or, in the more populated systems, even hundreds of galaxies orbit at velocities of several hundred kilometers a second within their common gravitational well, created by the overall mass which may be as great as a million billion solar masses. The remaining gas that has not been incorporated into individual galaxies has also reached equilibrium. Now, the temperature of gas in gravitational equilibrium is directly related to the total mass responsible for the gravitation. The mass of clusters of galaxies (typically 10^{14}–10^{15} solar masses) is such that the gas may reach temperatures of 10-100 million degrees. It will then emit X-rays.

Ever more efficient instruments have enabled us to deduce directly, at least for nearer systems, the mass M_X of the (baryonic) gas responsible for the observed X-radiation, the hypothesis being that the hot gas is in equilibrium, like the gas present in a star. This involves measurements of, on the one hand, the temperature of the gas according to the nature of the X-ray spectrum, and on the other, the total X-ray luminosity, which, assuming the gas to be in equilibrium, will readily give a value for M_X. We can also determine the total mass of the cluster in question from the observed velocities of its galaxies. These velocities are themselves reflections of the gravitational well within which they move. This dynamical mass is, as has already been stated in the case of galaxies, much greater than the luminous mass of the cluster's member galaxies. The situation thus resembles that for spiral galaxies: it is necessary to invoke the presence of a certain amount of dark matter to account for the observations.

It might have been thought that this X-ray gas, only recently detected, indeed accounted for the missing mass in the clusters: but this is not the case. The mass of X-ray gas is not sufficient to constitute the hidden mass, and the role has consequently been ascribed to dark matter. However, to the astronomers' surprise, this mass M_X is far greater (by 5 to 10 times) than the total mass of the galaxies within clusters. So clusters of galaxies are in a way not well named, though, in the final analysis, the estimate of their contribution to the baryonic ensemble (Table 3.1 and Figure 3.9) has risen to approximately 0.2 percent of the total cosmic energy/matter content. This means that the gas within the clusters (themselves comparatively rare objects, and certainly much fewer in number than lone galaxies)

Figure 3.8 These two images show the galaxy cluster Abell 85, located about 740 million light years from Earth. The composite image (facing page) shows, as a purple glow, the emission from multi-million degree gas detected in X-rays by NASA's Chandra X-ray Observatory and the other colors show galaxies in an optical image from the Sloan Digital Sky Survey (shown above). This galaxy cluster is one of 86 observed by Chandra to trace how dark energy has stifled the growth of these

massive structures over the last 7 billion years. Galaxy clusters are the largest collapsed objects in the Universe and are ideal for studying the properties of dark energy, the mysterious form of repulsive gravity that is driving the accelerated expansion of the Universe (X-ray (NASA/CXC/SAO/A.Vikhlinin et al.); Optical (SDSS)).

Table 3.1 Fractions of ordinary (or baryonic) matter existing as stars and gas within galaxies, and hot gas emitting X-rays within clusters, compared with total energy/matter in the cosmos, expressed as different parameters of Ω_b.

Nature of ordinary matter	Contribution to total energy/matter content of universe	
Hot gas emitting X-rays	$\Omega_{b\text{-gas}}$	~0.2%
Stars and gas in galaxies	$\Omega_{b\text{-stars/gas}}$	~0.32%
Total		**~0.52%**

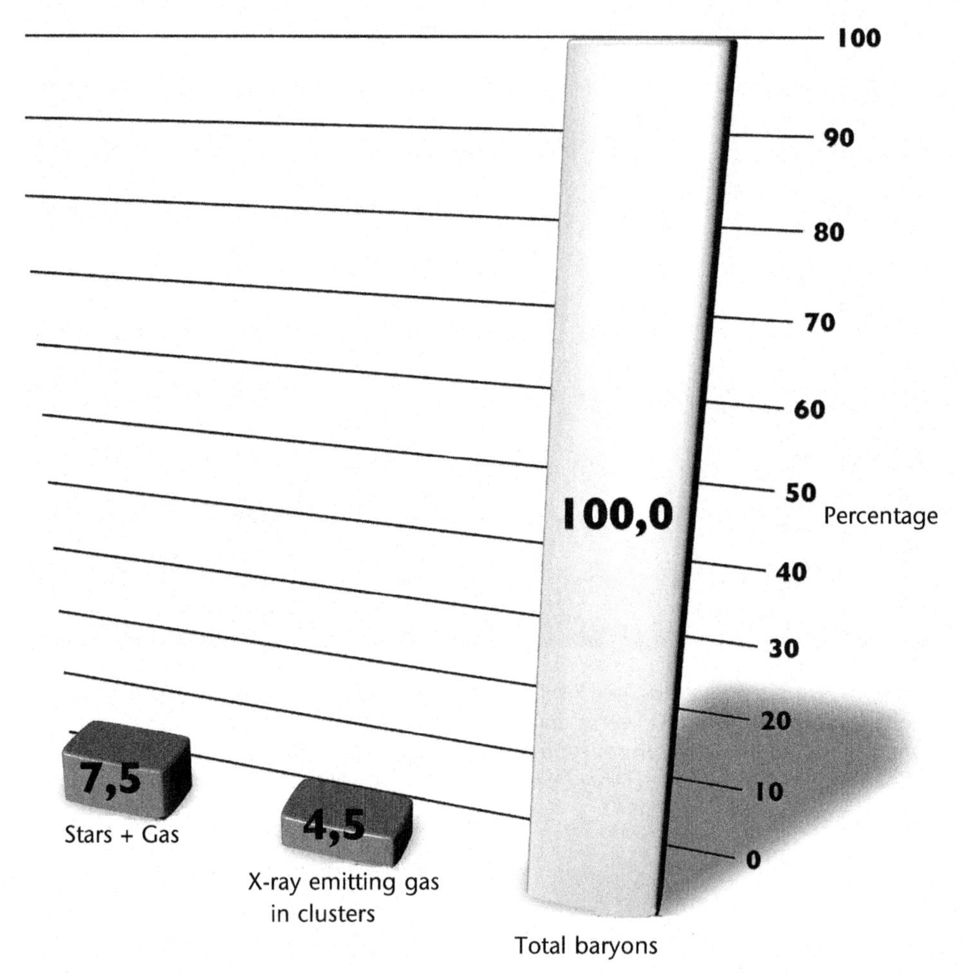

Figure 3.9 Comparisons of quantities of ordinary matter detected as stars and gas within galaxies, and as hot gas emitting X-rays within clusters, with the total content of this baryonic matter in the cosmos. An important fraction of the baryons is still missing from the total.

contributes almost as much as the stars in all the galaxies in the universe (see chapter 2).

Hot or cold: no easy reckoning

'Weighing' the baryons of even the nearby universe is, as we have seen, no easy task. Since their formation at the very beginning of the universe, these baryons have had varied careers. Some have come together to form stars, which themselves have become members of the galaxies so magnificently portrayed to us by the Hubble Space Telescope. Others have escaped this small-scale condensation process and gone on to populate the largest structures in the universe, clusters of galaxies, bathing them in a very tenuous envelope (just a few particles per cm^3), raised to a temperature of millions of degrees by the intense gravitational fields of these very massive objects. Stars within galaxies usually reveal their presence through their visible light, although less massive stars or those veiled by dust have to be detected in the infrared. The inventory of the 'stellar' content of these galaxies can only be complete if we take into account this dust ('visible' in the infrared) and the gas (neutral hydrogen) accessible to radio telescopes. The gas inside the clusters, not being directly visible, has long been hidden from us. In the twentieth century, as spaceflight developed, its X-ray emissions were finally revealed. It can do little to boost our deflated egos, knowing that the contents of the cosmos are in the main unseen, that X-ray baryons contribute only about as much as their stellar counterparts. Nevertheless, we can at last try to estimate, for the local universe, the quantity of ordinary matter detected in the various forms so far revealed to us. The provisional tally of our search for baryons, taking into account the estimates already mentioned for stars and the much more marginal estimates for dust and molecular gas, then adding the hot gas of the clusters, leads us to a total value of about 0.5 percent.

We see at once from Table 3.1 and Figure 3.9 that there is still much left to be found, in comparison to both the global content expressed as $\Omega_{tot}= 1$ and to the total of baryons. It seems then that matter indeed tends to be dark... But, then again, are these estimates correct? Is there not some other means of working out the total quantity of baryons in the universe without having to detect them in each of their various guises?

This will be discussed in the next chapter.

4 Cosmic Cluedo: where, when and how?

"I keep six honest serving-men
(They taught me all I knew);
Their names are What and Why and When
And How and Where and Who."

Rudyard Kipling

The observation of the recession of the galaxies by Edwin Hubble, Vesto Slipher and Milton Humason, who showed that all the galaxies they were observing (some tens of galaxies at that time) were moving away from the Milky Way, is a reflection of the overall expansion of the universe with time (Figure 4.1). The effect of this expansion is the inexorable dilution of the energy/matter content of the universe. The density and temperature of the various components will continue to decrease with time (with the exception of certain categories of dark energy, whose density remains constant, in the manner of the cosmological constant[1]). If we mentally reverse this process, we immediately arrive at the idea, formally elaborated by physicists Georges Lemaître and, later, George Gamow (Figure 4.2), of a universe increasingly hotter and more dense in the past. In the classic vision where we rewind time to zero, we achieve the now well known idea of an initial singularity, and a 'Big Bang'.

A more complete vision shows, however, that, given the current state of our knowledge, we can go back no further than the Planck time.[2] This concept of an initially hot, dense universe is borne out by, among other

---

[1] The cosmological constant (Λ) was a term introduced by Einstein into his cosmological equations, which enabled him to find solutions based on a static universe. The discovery of the recession of the galaxies by Hubble, and of cosmic expansion, cast doubt upon this constant (Λ). Cast aside for decades, it was resurrected as a possible cause of the acceleration of cosmic expansion. It has thus become a candidate for the famed dark energy responsible for this acceleration.

[2] Our current knowledge cannot give us insight into what occurred before the Planck time ($\sim 10^{-43}$s, see Appendices). This involves a blending of General Relativity and quantum mechanics, an objective still to be attained by theoretical physics.

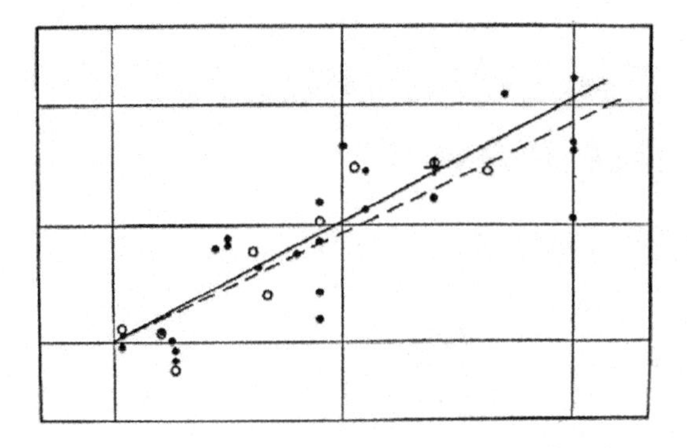

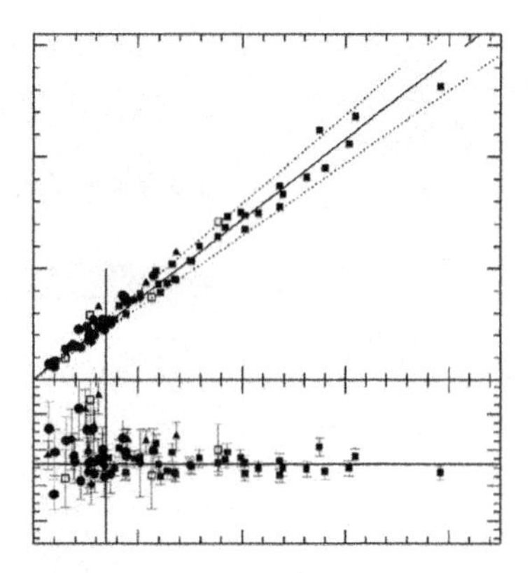

Figure 4.1 (Left) A graph, by Edwin Hubble himself, demonstrating the relationship between the velocity of recession of 'runaway' galaxies and their distance in parsecs (1 parsec = 3.26 light years). (Right) Recent estimates of this relationship over distances 400 times greater than those of Hubble's era. The result is unequivocal: the farther away a galaxy is, the faster it is receding from the observer. The value originally allotted to Hubble's constant of proportionality, H_0, was 500 km/s/Mpc (500 kilometers per second per million parsecs from the observer). Current measurements suggest a value of 72 km/s/Mpc. The determination of the constant H_0 was the subject of many controversial debates, engendering numerous revisions. These arose from the difficulty of 'surveying' the universe step-by-step using different methods of distance determination according to the scale involved.

observations, the detection of the cosmic microwave background (see chapter 5). Photons from this 'fossil' radiation, as detected by the COBE and WMAP satellites, are at a temperature of about 3 degrees Kelvin. They have their origins in the final interactions with matter when the universe was about 300,000 years old (i.e. at redshift z = 1,000). This matter, previously ionized (i.e. with electrons and nuclei separated), recombined into neutral atoms at a temperature of about 3,000 degrees. Photons were liberated, for there were now no isolated electrons to participate in games of 'cosmic billiards' with them, and their temperatures subsequently fell from 3,000 K to 3 K, simply as a result of the expansion.

We can therefore use the cosmological model to reconstruct the thermal history of the cosmos, on the basis of the evolution of temperature and density through time, employing the tools of high-energy physics, capable of describing very hot and very dense environments (Figure 4.3 and Appendices). Thus equipped, we can at least sketch out the broad picture in answer to those questions about the origin of matter posed at the beginning of this chapter.

Figure 4.2 Some important figures in the 'cosmological saga'. (Upper left) Georges Lemaître, long unacknowledged 'father' of the Big Bang, with his theory of the 'primitive atom'. (Upper right) George Gamow, contributor to the elaboration of the Big Bang model by writing, with Ralph Alpher and Hans Bethe, the original article on the formation of elements in the universe and prediction of the cosmic background radiation (American Institute of Physics). (Below, left to right) Albert Einstein, Edwin Hubble and Walter Adams at the Mount Wilson Observatory (Archives of the California Institute of Technology).

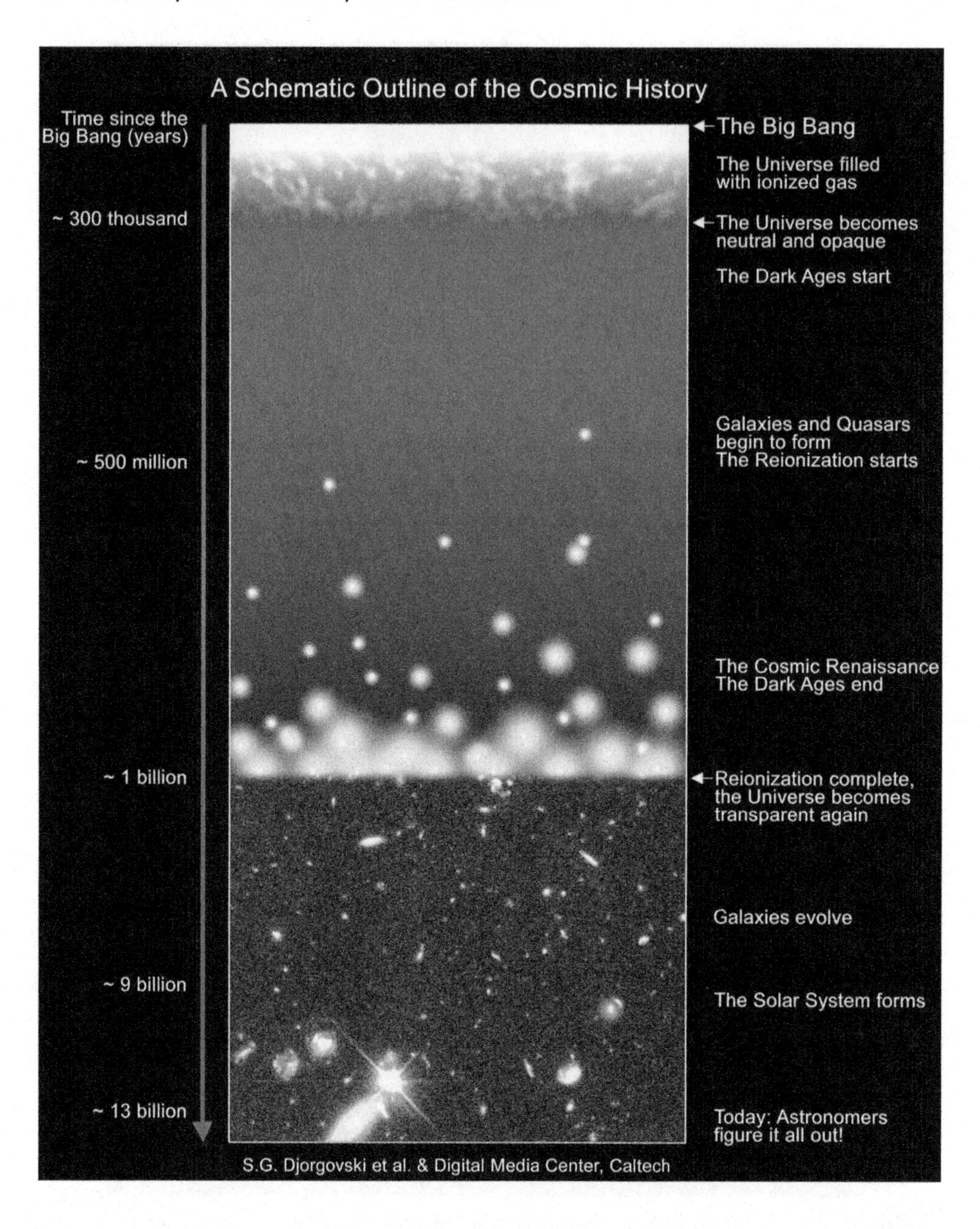
A Schematic Outline of the Cosmic History
Time since the Big Bang (years)
~ 300 thousand
~ 500 million
~ 1 billion
~ 9 billion
~ 13 billion
The Big Bang
The Universe filled with ionized gas
The Universe becomes neutral and opaque
The Dark Ages start
Galaxies and Quasars begin to form
The Reionization starts
The Cosmic Renaissance
The Dark Ages end
Reionization complete, the Universe becomes transparent again
Galaxies evolve
The Solar System forms
Today: Astronomers figure it all out!
S.G. Djorgovski et al. & Digital Media Center, Caltech

A creative youth

The first moments after the Big Bang saw the genesis of the various particles which would go on to populate the universe and would lead to the kind of matter with which we are familiar: the constituents of atoms, together with neutrinos and dark matter. The detailed description of this genesis is complex, but can, thanks to Einstein, be summed up by the now famous formula,

$$E = mc^2,$$

which describes the equivalence of mass and energy. This formula is now universally known because it is the basis of the transformation of atomic energy into electricity, and also of the workings of the atomic bomb. The essential thing to bear in mind here is that from mass it is possible to produce energy, and *vice versa*. This reciprocity applies in theory at all times in cosmic history, and we can therefore write:

$$\text{energy} \rightarrow \text{particle} + \text{anti-particle}$$

always remembering, of course, that the supply of energy has to be sufficient for the mass of the particle in question (especially since an anti-particle is created at the same time as the particle) – in other words, provided that the universe is hot enough.

One of the virtues of the cosmological model is that it predicts the variation in temperature, expressed also in terms of energy, as a function of the cosmic time elapsed since the Big Bang. So, at a cosmic age of one ten-

Figure 4.3 The universe started with the Big Bang around 13.7 billion years ago, and from the Planck time (10^{-43} seconds) onwards its history is generally characterized by two great eras. First, there was an era during which the energy-matter content of the universe was dominated by radiation (the Radiation-Dominated Era), which ended when the Universe was about 70,000 years old. Second, there was the Matter-Dominated era, during which it was the material content that prevailed.. Eventually, by around 300,000 years after the Big Bang, the temperature had dropped to about 3000 K, and atomic nuclei and electrons had combined to make atoms of neutral gas. The glow of this 'Recombination Era' is now observed as the cosmic microwave background radiation. The universe then entered the 'Dark Ages', which lasted for about half a billion years, until they were ended by the formation of the first galaxies and quasars. The light from these new objects turned the opaque gas filling the universe into a transparent state again, by splitting the atoms of hydrogen into free electrons and protons. This Cosmic Renaissance is also referred to by cosmologists as the 'Reionization Era', and it signals the birth of the first galaxies in the early universe (S.G. Djorgovski *et al.*, Caltech and the Caltech Digital Media Center)

thousandth (10^{-4}) of a second, we calculate the temperature to be a thousand billion degrees.[3]

At one second, this has fallen to ten billion degrees, and thereafter we divide the temperature by ten as the time is multiplied by 100. The various particles existing in Nature may therefore be formed when the temperature and energy available correspond to their mass. Any given particle cannot afford to miss its chance, since the temperature of the universe will never be as high again, except in the cores of stars where temperatures are of the order of many millions of degrees. If we go back far enough in time, the temperature – beyond ten thousand billion degrees (1 GeV) – is such that the cosmic fluid is still a plasma[4] of quarks and gluons (which the Large Hadron Collider in Geneva has recently succeeded in recreating). This means that the quarks, the basic building blocks of atomic nuclei, have not yet come together to form hadrons (protons and neutrons). When the temperature falls to approximately the 1 GeV level, the quarks will finally be incorporated into protons and neutrons. Ordinary matter such as that which now fills the universe will have been created (see Appendices).

All other particles were also created during this process, leaving the universe globally neutral with as many protons as electrons. Neutrinos, also formed during the first moments of the universe, interact very weakly with other particles. They decoupled from the rest of the cosmic fluid at a temperature of ten billion degrees, and henceforth constitute a cosmic background radiation similar to that of photons, created at the time of the recombination (see chapter 5). The direct detection of this 'bath' of neutrinos cannot currently be undertaken because of their very low energy and the fact that they exhibit little interaction (there are approximately 150 neutrinos per cm^3, as opposed to 400 in the case of the photons of the cosmic microwave background radiation, discovered by Penzias and Wilson, in the normal environment). Ever more accurate measurements of the characteristics of the cosmic background radiation reveal their presence.

---

[3] $T = 10^{12}$ K, i.e. an energy of 100 MeV (see Appendix on temperature-energy transformations).

[4] A plasma is a gas that is globally neutral, but its electrons are not yet associated with the nuclei (i.e. it is ionized), usually because of very high temperature or exposure to very energetic radiation. Plasma is considered to be the fourth state of matter. The interior of stars consists of plasma at temperatures of many millions of degrees. Before a cosmic age of 300,000 years, the cosmic fluid was also in the form of plasma, and would be so again after the so-called 'epoch of reionization'.

Where do dark matter and anti-matter fit in?

Neutrinos were very much in the spotlight during the 1980s. Apart from the fact that the neutrino was already a well known particle (predicted by Wolfgang Pauli in 1930 and discovered in 1956; there are three families of neutrinos), measurements taken during this decade indicated that the neutrino, hitherto thought to be massless, in fact did have mass. Moreover, this mass was considerable enough to have a cosmological impact. It seemed that dark matter had at last been identified.

Alas, measurements of the mass of neutrinos were revised downwards, and because of their very low mass, neutrinos were finally seen as playing only a minor role. Other candidates would have to be found to account for the 'missing' mass. Fortunately for astrophysicists, the extensions of the standard model of particle physics, such as super-symmetry theories, are good enough to come up with new (super-) particles. In fact, there were now large numbers of particles bearing the suffix '-ino'. Without going into too much detail here, pride of place among these particles goes to a neutral, non-interacting particle which will not spontaneously disintegrate. This describes the neutralino,[5] with a mass between several tens of GeV and several TeV. This dark matter, whatever it may be, is also fervently sought.

The attentive reader will have noticed that from a given quantity of energy, there may be created not just a particle but also its associated anti-particle. Consequently, during the thermal history of the universe, matter and anti-matter were formed in equal quantities. Anti-particles have the same mass as their 'mirror' particles, but they are of opposite quantum number, especially as regards their electric charge. When a particle of matter meets its anti-particle, they mutually annihilate, releasing all their energy (i.e. the sum of their kinetic energy and the energy corresponding to their mass), reversing the reaction which led to their formation (Figure 4.4). When they mutually annihilate, a particle and its anti-particle are transformed into a shower of other particles, principally photons.

According to the standard scheme of particle physics as applied to cosmology, baryons and anti-baryons, having been formed, rapidly annihilated each other, with the exception of a few residual 'islands'.[6] However, in this context, the prediction for the quantity of matter surviving this phenomenon of annihilation is far lower (by several orders of magnitude)

---

[5] The neutralino is a light particle, possibly the lightest, predicted (though still hypothetical) by extensions of the standard model of particle physics. The neutralino is the favored candidate for dark matter in the universe.

[6] If this had not occurred, there would be no cosmologists and physicists around to contemplate these things!

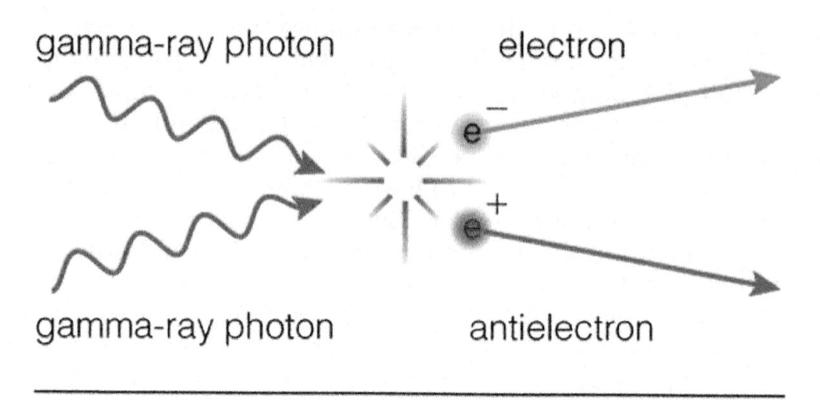

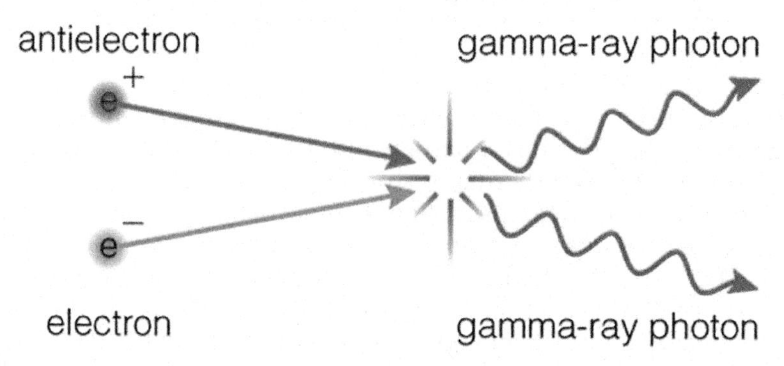

Figure 4.4 Matter-anti-matter annihilation. In the early universe, matter is created via pair production and destroyed by annihilation. Since an equal amount of matter and anti-matter is produced in pair production, there could be a problem. *Pair Production*: to make pairs, the temperature must be high so that photons (or whatever) can make virtual particles real, and since the Universe cools as it expands, then at some point a threshold is reached beyond which pairs can no longer be created. *Annihilations*: annihilations don't require high temperatures. All that is needed is a density that is sufficiently high so that the probability that a matter particle and its anti-matter twin will re-unite is high. In the early Universe, this allows the annihilation reactions to persist long after pair production halts. Because of the symmetry of pair production and annihilation, calculations suggest that annihilations would remove all of the matter/anti-matter from the Universe leading to a Universe which is filled entirely with photons. Obviously this is not true since we are here! The problem is more than a little vexing because the asymmetry is small. There must have been 1,000,000,001 electrons for every 1,000,000,000 positrons in the early Universe. If an asymmetry at this level did not exist, then we would not be here. This very small matter/anti-matter asymmetry is another mystery of the universe which we must try to explain.

than astronomical observations suggest. Also, we should be observing radiation emitted at the boundary between zones where matter and anti-matter meet, and this is not the case. The universe appears completely empty of anti-matter (see chapter 9).

There is clearly a difficulty, which seems to arise from the standard model of particle physics. The model therefore requires modification. For this reason, physicist Andrei Sakharov invoked the possibility that certain principles of the standard model could be violated, leading to asymmetry in the production of baryons at the expense of anti-baryons. Certain recent theoretical extensions of the standard model, for example super-symmetry, 'naturally' violate the principles of the standard model in question and could explain the absence of anti-matter, because certain reactions are in disequilibrium, with 100,000,001 particles of matter formed for every 100,000,000 particles of anti-matter!

The alchemy of the first three minutes

Thermal history proceeds. As we shall see below, around the first three minutes, as the temperature fell from ten billion to one billion degrees, protons and neutrons, formed a few moments before, would combine to synthesize the nuclei of the lightest elements (primordial or Big Bang nucleosynthesis); heavier elements would be produced many millions or billions of years later inside stars (stellar nucleosynthesis). At a temperature of 10,000 degrees, the density of matter (ordinary and dark), became equal to that of the radiation. Immediately afterwards, there occurred the recombination during which atoms were formed and the process of the formation of large structures in the universe began, dominated by dark matter. Everything was in place for the formation of galaxies, stars, planets and life

Once individual particles had been created, their destinies remained uncertain. They would in effect participate in various processes via fundamental interactions to which their natures made them susceptible (electromagnetic, strong, weak, gravitational), all the while undergoing the implacable dilution due to cosmic expansion. It is therefore easy to comprehend that interactions between particles could only happen within certain very precise 'windows' in time. If the cosmic fluid is too dilute, the distance between particles becomes greater than the length at which the physical interaction operates efficiently. This general principle also applied to the epoch during which the first chemical elements were formed.

The origin of the totality of the elements of which the universe is composed, now brought together in the famous periodic table created by Mendeleev (Figure 0.2 in the Introduction), remained a mystery for a very long time. In the late 1940s Gamow, realizing that an expanding universe must have been at some time very hot and dense, put forward the idea that thermonuclear reactions must indeed have taken place there. He presented

PHYSICAL REVIEW VOLUME 73, NUMBER 7 APRIL 1, 1948

Letters to the Editor

PUBLICATION of brief reports of important discoveries in physics may be secured by addressing them to this department. The closing date for this department is five weeks prior to the date of issue. No proof will be sent to the authors. The Board of Editors does not hold itself responsible for the opinions expressed by the correspondents. Communications should not exceed 600 words in length.

The Origin of Chemical Elements

R. A. ALPHER*
Applied Physics Laboratory, The Johns Hopkins University, Silver Spring, Maryland
AND
H. BETHE
Cornell University, Ithaca, New York
AND
G. GAMOW
The George Washington University, Washington, D. C.
February 18, 1948

As pointed out by one of us,[1] various nuclear species must have originated not as the result of an equilibrium corresponding to a certain temperature and density, but rather as a consequence of a continuous building-up process arrested by a rapid expansion and cooling of the primordial matter. According to this picture, we must imagine the early stage of matter as a highly compressed neutron gas (overheated neutral nuclear fluid) which started decaying into protons and electrons when the gas pressure fell down as the result of universal expansion. The radiative capture of the still remaining neutrons by the newly formed protons must have led first to the formation of deuterium nuclei, and the subsequent neutron captures resulted in the building up of heavier and heavier nuclei. It must be remembered that, due to the comparatively short time allowed for this process,[1] the building up of heavier nuclei must have proceeded just above the upper fringe of the stable elements (short-lived Fermi elements), and the present frequency distribution of various atomic species was attained only somewhat later as the result of adjustment of their electric charges by β-decay.

Thus the observed slope of the abundance curve must not be related to the temperature of the original neutron gas, but rather to the time period permitted by the expansion process. Also, the individual abundances of various nuclear species must depend not so much on their intrinsic stabilities (mass defects) as on the values of their neutron capture cross sections. The equations governing such a building-up process apparently can be written in the form:

$$\frac{dn_i}{dt}=f(t)(\sigma_{i-1}n_{i-1}-\sigma_i n_i) \quad i=1,2,\cdots 238, \tag{1}$$

where n_i and σ_i are the relative numbers and capture cross sections for the nuclei of atomic weight i, and where $f(t)$ is a factor characterizing the decrease of the density with time.

We may remark at first that the building-up process was apparently completed when the temperature of the neutron gas was still rather high, since otherwise the observed abundances would have been strongly affected by the resonances in the region of the slow neutrons. According to Hughes,[2] the neutron capture cross sections of various elements (for neutron energies of about 1 Mev) increase exponentially with atomic number halfway up the periodic system, remaining approximately constant for heavier elements.

Using these cross sections, one finds by integrating Eqs. (1) as shown in Fig. 1 that the relative abundances of various nuclear species decrease rapidly for the lighter elements and remain approximately constant for the elements heavier than silver. In order to fit the calculated curve with the observed abundances[3] it is necessary to assume the integral of $\rho_n dt$ during the building-up period is equal to 5×10^4 g sec./cm^3.

On the other hand, according to the relativistic theory of the expanding universe[4] the density dependence on time is given by $\rho\cong10^6/t^2$. Since the integral of this expression diverges at $t=0$, it is necessary to assume that the building-up process began at a certain time t_0, satisfying the relation:

$$\int_{t_0}^{\infty}(10^6/t^2)dt\cong5\times10^4, \tag{2}$$

which gives us $t_0\cong20$ sec. and $\rho_0\cong2.5\times10^3$ g sec./cm^3. This result may have two meanings: (a) for the higher densities existing prior to that time the temperature of the neutron gas was so high that no aggregation was taking place, (b) the density of the universe never exceeded the value 2.5×10^3 g sec./cm^3 which can possibly be understood if we

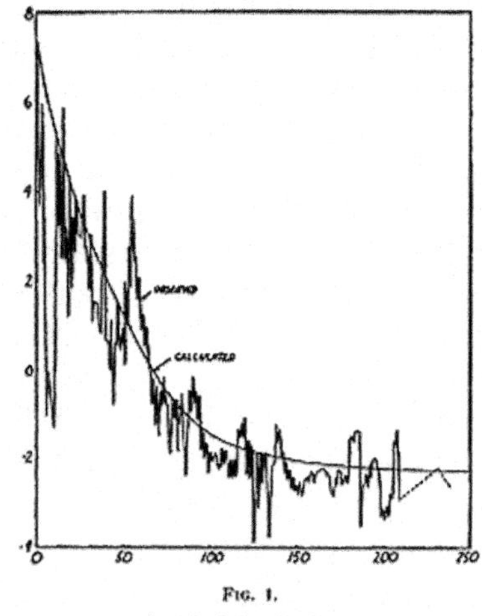

FIG. 1.
Log of relative abundance
Atomic weight

Figure 4.5 Facsimile of the original 'Alpha-Beta-Gamma' article. Entitled simply 'The Origin of Chemical Elements' this groundbreaking paper was authored by Ralph Alpher, Hans Bethe and George Gamow, and published in *Physical Review Letters*, Vol. 73, No. 7, 1 April 1948.

his results in an article written with his student Ralph Alpher, which became well known for two reasons: on the one hand, it was a brilliant scientific article, and on the other hand because Gamow added the name of physicist Hans Bethe in order to create the pleasing effect of its authors being 'Alpha', 'Beta' and 'Gamma' (Figure 4.5).

In Gamow's view, most of the heavy elements would have been synthesized during the first moments of the universe. From this hypothesis involving a very hot phase, the famous physicist deduced the probable existence of a radiation filling the whole universe. However, it was soon realized that this scenario would not be able to explain the formation of elements heavier than beryllium. Only later was it understood that heavy elements are created in the nuclear reactors that are the cores of stars. Conversely, these nuclear 'factories' cannot have formed all the elements. For example, deuterium,[7] an isotope of hydrogen which became famous for its role in the 'heavy water' battle during the Second World War, is a very fragile element, easily destroyed in the stellar environment. Since we observe deuterium in the universe, it must have been formed in some other place, i.e. the primordial universe. The period during which the so-called light elements, such as hydrogen and helium, were formed has therefore been called the era of primordial or Big Bang nucleosynthesis (BBN), extending from 3-20 minutes after the Big Bang, and the heavy elements many millions or billions of years later within stars, by stellar nucleosynthesis.

When protons and neutrons get together

Protons and neutrons are the building blocks of atomic nuclei, and are held together within them by the so-called strong nuclear force. As long as the temperature remains above ten billion degrees, nuclei will be dissociated into protons and neutrons because their agitation energy is greater than the energy of the strong interaction which tends to pull them together. It is therefore necessary for the temperature to be lower so that the nuclear force can overcome the agitation. It is also necessary that the environment should not be

---

[7] Deuterium (symbol: D) is an isotope of hydrogen with a nucleus formed of one proton and one neutron. Easily destroyed in nuclear reactions, the deuterium formed during primordial nucleosynthesis is nowadays difficult to detect. (Isotopes are elements having the same atomic number, i.e. the same number of protons, but whose nuclei have a different number of neutrons.) Deuterium oxide (D_2O), unlike normal water which is hydrogen oxide, is heavier (hence the name 'heavy water') since its nucleus contains one neutron as well as a proton, and it is effective in slowing neutrons within nuclear reactors.

too dilute, or the distances between particles will be greater than the influence of the force of fusion within the nuclei which, contrary to gravitation or electromagnetic interaction, can act only over extremely short distances, of the order of the size of atomic nuclei. Finally, unlike protons, isolated neutrons are unstable particles. They transform into protons *via* the weak interaction. Their lifetime is only about ten minutes, a period beyond which the population of neutrons will have disappeared. We therefore see that the necessary conditions for the marriage of protons and neutrons to occur are very restricted. The date for this 'wedding' is fixed: everything has to occur within the first few minutes of the universe's existence. During this short period, the temperature drops below ten billion degrees and the density is right for the strong interaction to occur and for enough neutrons to be present to trigger the interactions. One proton and one neutron can finally form a deuterium nucleus which, although it is fragile, can resist the destructive force of the great numbers of ambient energetic photons.

All the baryons

The dice are cast. Now, a series of fusion reactions is triggered, in the course of which are formed helium, lithium, tritium, beryllium and their various isotopes. The resulting comparative abundance by mass is of the order of 75 percent hydrogen and 25 percent helium, the contribution of the remaining elements, although present, being extremely insignificant. All this is in agreement with observations, and is one of the great success stories of the Big Bang model. Contrary to other more speculative aspects of the cosmological model, (nuclear) physics, which is brought to bear to describe this phase of the synthesis of the elements, is a well established branch of physics and does not depend upon poorly understood parameters. The proportions of the different elements formed are therefore related, unquestionably and not on the basis of some unclear model, to the density of the environment and therefore to the total quantity of baryons.

This serves only to reinforce the fundamental conclusion: we know the *total baryonic content* of the universe with great accuracy. It is baryons that populate the universe we observe. Only their relative abundance might evolve as the universe becomes more diluted. We should find the same content for all cosmic epochs. The verdict is therefore unalterable. The density of baryons resulting from the primordial nucleosynthesis, which agrees extremely well with the observed proportions, a density corrected for the expansion of the universe, leads to the current value for the density parameter $\Omega_{b\text{-}tot}$

$$\Omega_{b\text{-}tot} \sim 4.4\%$$

So, as we see, the laws of nuclear physics, applied to the primordial universe,

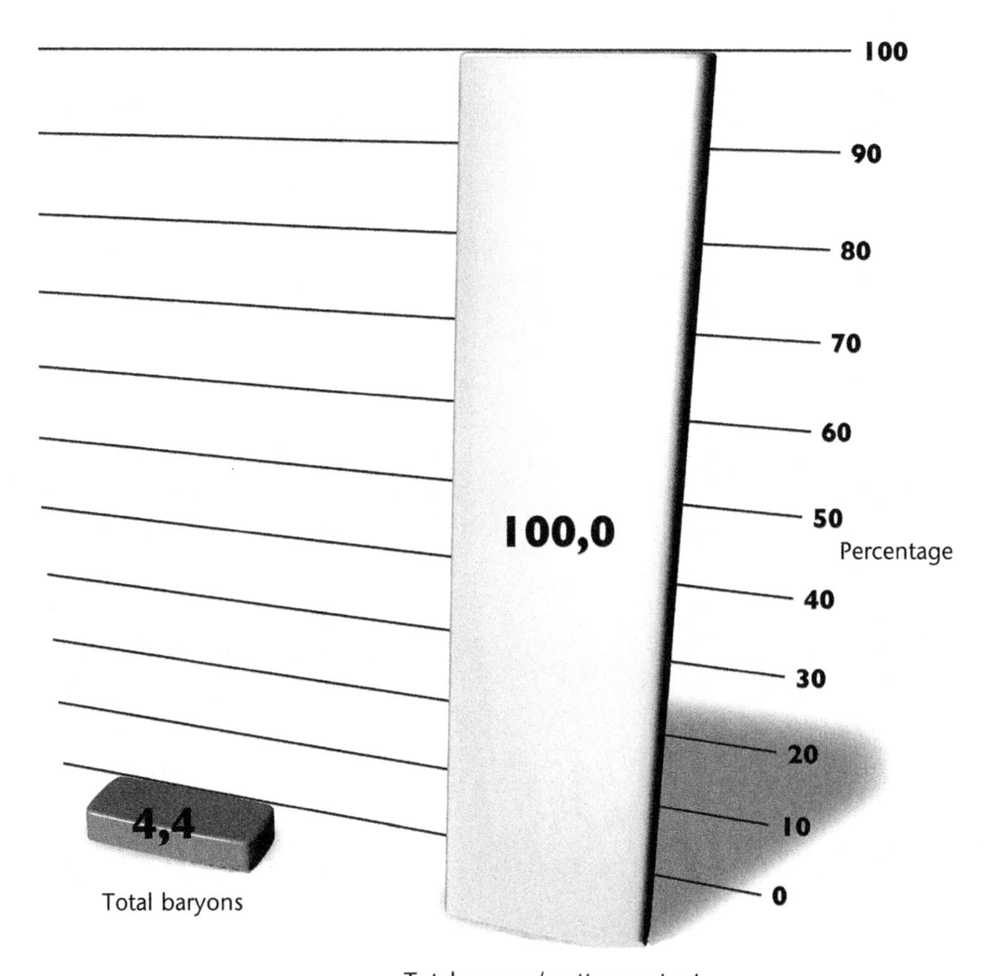

Figure 4.6 The fraction of all the baryons (all the ordinary matter) of the cosmos in relation to its total energy/matter content. Ordinary matter represents only a tiny part compared with dark matter and dark energy.

tell us unambiguously that the total of *all* the baryons contained within it represent only 4 to 5 percent of the content necessary to explain the expansion as measured (Figure 4.6).

Their contribution is certainly a poor one, but what is really frustrating is that we had not even found all those baryons during our previous searches!

5 300,000 years on: all present and correct

"Keep your accounts, keep your friends." French proverb

Let us take up our story again, travel to a time around 300,000 years after the Big Bang, and imagine two great moments, both crucial for the history of the universe.

Matter takes over

Almost in tandem two historical transitions occurred around this time. One is known as the epoch of recombination, the moment when atomic nuclei and electrons combined to form atoms. The other was the moment when the era of matter began, instantly dominating radiation in terms of energy density. As we have seen, the expansion of the universe implies, as we wind back through time, the existence of very hot and very dense phases in its earliest stages. We have also seen how, given such physical conditions during those famous first three minutes, the totality of particles appeared with, in particular, the formation of light elements. After the first three minutes, the universe consisted of a mixture of matter (mostly dark) and photons. As cosmic expansion proceeded, there was an inexorable dilution of the energy/matter content. However, this dilution did not affect matter and radiation in the same way. In the case of matter formed of particles, the density (the number per unit of volume) decreased as the volume increased. In the case of photons, the particles constituting radiation, there was a decrease not only in their density, in the same way with ordinary particles, but also in their energy, owing to cosmological redshift. These two effects combined to cause the energy contained in radiation to decline more rapidly than that in matter.

The radiation that had hitherto dominated the energy density was gradually overtaken by matter. There came a time, known as the epoch of matter-radiation equality, after which matter dominated the universe (Figure 5.1). It can be easily shown that this occurred at redshift z_{eq}, which can be expressed as a function of the values for the density of matter and radiation in the present universe.[1]

---

[1] The relationship is $1 + z_{eq} = [\Omega_m/\Omega_R]_0$, where the zero represents the current epoch.

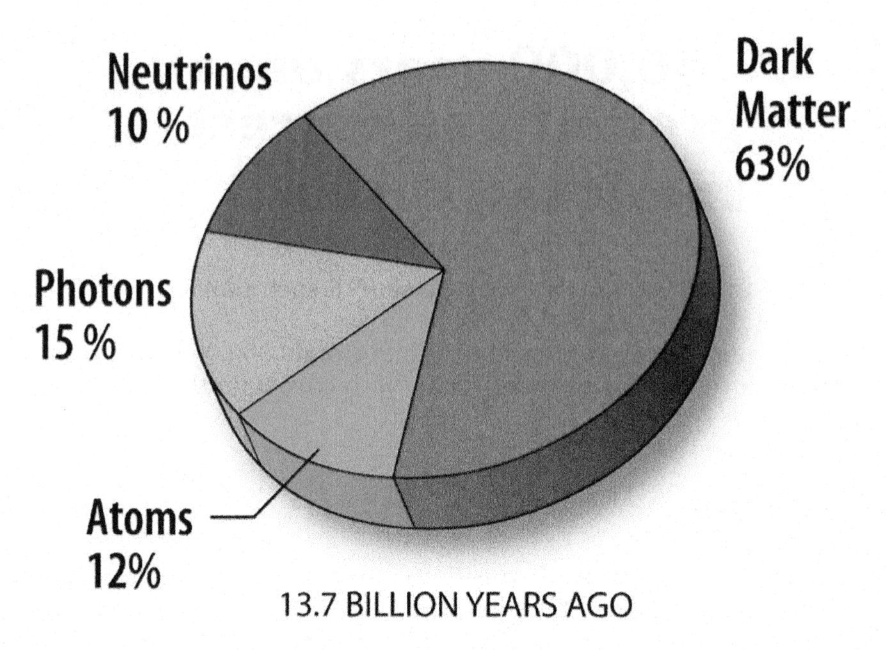

Figure 5.1 By studying the cosmic microwave background radiation seen by the Wilkinson Microwave Anisotropy Probe (WMAP) from a time when the universe was only about 300,000 years old ($z = 1,200$), the energy/matter content of the universe at that time has been revealed as shown in this pie chart. At $z = 1,200$, matter (mainly dark) already dominates radiation, with neutrinos making up 10 percent of the universe, atoms 12 percent, dark matter 63 percent, photons 15 percent, and dark energy was negligible. In contrast, estimates from WMAP data show the current universe consists of 4.6 percent atoms, 23 percent dark matter, 72 percent dark energy and less than 1 percent neutrinos (see Figure 5.7) (NASA/WMAP Science Team).

Using the values as measured by the Wilkinson Microwave Anisotropy Probe (WMAP) for the density parameters Ω_m and Ω_R, we obtain $z_{eq} = 3,000$, which corresponds to an age for the universe of approximately 60,000 years after the Big Bang. Before this epoch, the temperature of the cosmological fluid remained high enough for electrons and protons, normally mutually attractive, to stay apart, unable to form hydrogen atoms because of their incessant agitation. But the temperature decreased inexorably, the agitation of the particles lessened and the formation of atoms through the marriage of electrons and protons became possible. This process did not, of course, occur instantaneously. At redshift $z = 1,200$ (equivalent to 300,000 years), about 90 percent of the electrons had already been incorporated into hydrogen atoms. This (re-)combination was also an electrical neutralization of the cosmic fluid, an occurrence that had immense consequences for the future of the cosmos and left an indelible trace in the space-time chronicle, to the great good fortune of cosmologists.

A fossil in the sky

As long as matter remained ionized, i.e. with protons separated from electrons, there was a very strong bond between electrons and photons, in a state of permanent interaction. It was not a case of 'photons here and baryons there'. Unlike dark matter, which by definition does not participate in electromagnetic interactions, baryonic matter cannot exist independently from radiation. However, all this applies only if the matter is in the form of plasma. The (re-)combination of electrons and charged nuclei into atoms spells an end to the cosmic billiard game, with its photons and charged particles ceaselessly interacting: game over!

Even though this phenomenon is also not instantaneous, we can ascribe a date to it, of redshift $z_{rec} = 1{,}100$, i.e. shortly after the beginning of the era of the dominance of matter. At a stroke, photons, their paths no longer deviated by charged particles, could travel freely and propagate throughout the expanding universe. At a given moment (assuming the phenomenon to be instantaneous), all the photons are involved in one final interaction, a last decoupling with ordinary matter, and after several billion years, they may reach the eye of a terrestrial observer as if they came from a surface aptly called the surface of last scattering.[2] If the universe is indeed as homogeneous and isotropic as postulated by the Cosmological Principle (see below), the photons must all decouple at the same temperature, that of recombination – the process of the formation of hydrogen and helium atoms at a temperature of $T \sim 3{,}000$ K. The so-called 'era of recombination' lies at redshift $z_{rec} \sim 1{,}000$.

Because they were in equilibrium with matter as a result of their ceaseless interactions, the energy distribution of the photons is that of a black body.[3] It is therefore to be expected that, if we turn an appropriate instrument to the

---

2 At the time of the recombination, all photons decoupled almost instantaneously from matter and were finally free to propagate through the cosmos. They appear to come from the surface of a sphere for all observers apparently at the centre of this sphere. This is true for wherever in the universe the observer is situated. This is the surface of last scattering.

3 A so called black body is a totally absorbent heated body. A good way to visualize this is by imagining an oven equipped with an opening to allow observation of the radiation in its interior. The hotter the body becomes, the shorter the wavelengths emitted as radiation ('white heat'), and *vice versa*. This wavelength λ (in cm) is related to the temperature T (in degrees K) by Wien's Law: $\lambda T = 0.29$. The 3 K cosmic microwave background emits black-body radiation in the millimetric waveband. The Sun, whose photosphere behaves as a black body at 5,300 K, therefore emits radiation at around 550 nm, which we see as yellow.

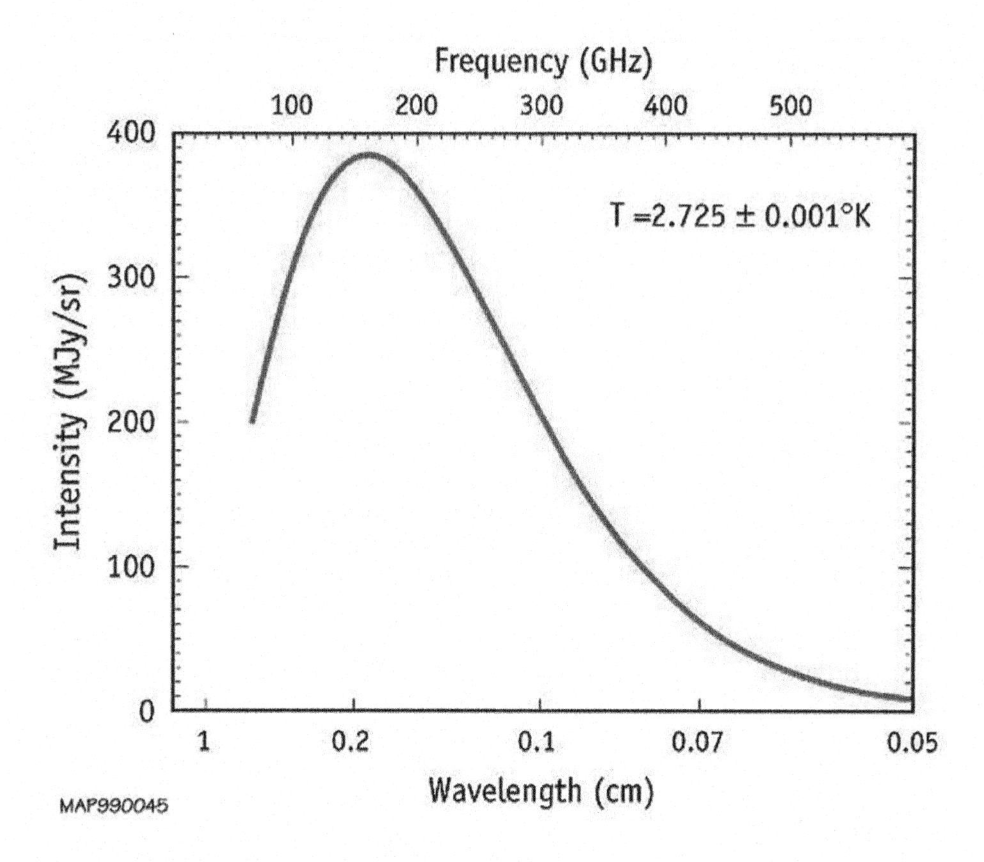

Figure 5.2 This figure shows the prediction of the Big Bang theory for the energy spectrum of the cosmic microwave background radiation compared with the observed energy spectrum. The FIRAS experiment on NASA's Cosmic Background Explorer (COBE) satellite measured the spectrum at 34 equally spaced points along the blackbody curve. The error bars on the data points are so small that they cannot be seen under the predicted curve in the figure! There is no alternative theory yet proposed that predicts this energy spectrum. The accurate measurement of its shape was another important test of the Big Bang theory (NASA/COBE Science Team).

sky, we observe, in all directions, radiation whose temperature has decreased, owing to the expansion of the universe, to its current value of T $\sim$3 K.

This is effectively what was discovered accidentally by Robert Wilson and Arno Penzias in 1965 as they experimented with the Bell Laboratories Horn Antenna at Crawford Hill, New Jersey (see Chapter 8). Their discovery was later confirmed by many other experiments, for example that carried out by the COBE satellite, which measured the spectrum of the cosmic microwave background, establishing that it was indeed that of a perfect black body, and

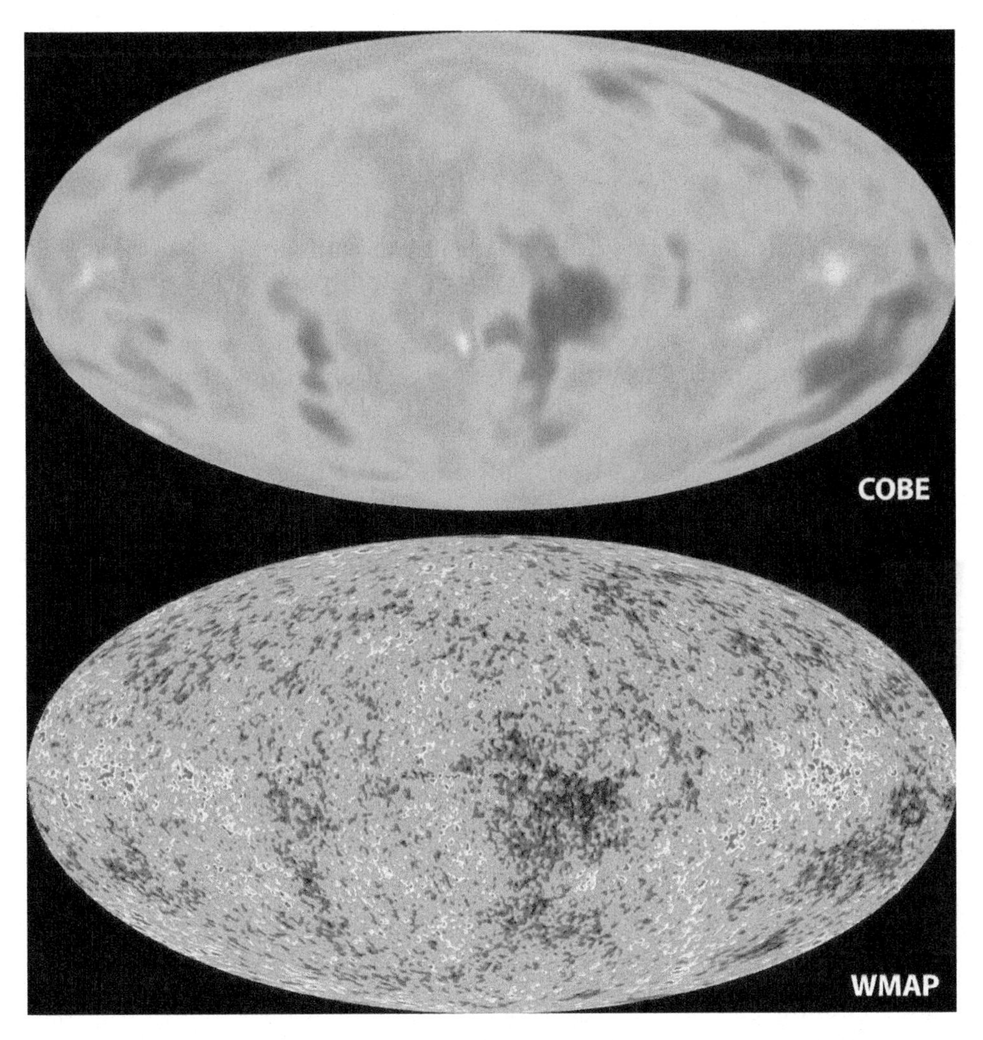

Figure 5.3 All-sky images of the infant Universe, about 300,000 years after the Big Bang, 13.7 billion years ago. In 1992, NASA's COBE mission (top image) first detected patterns in the oldest light in the universe (shown as color variations). WMAP brings the COBE picture into sharp focus (bottom image). The features are consistent and 35 times more detailed than COBE's, The new, detailed image provides firm answers to age-old questions (NASA/WMAP Science Team).

producing an all-sky temperature chart, a result which was improved very considerably by the WMAP spacecraft (Figures 5.2 and 5.3).

So the sky has a uniform temperature, a fact that validates the Cosmological Principle. Well, *almost* uniform If we study Figure 5.3 more closely, tiny variations in temperature appear, of about one part in a hundred thousand. Although they are infinitesimal, we shall see the importance of these fluctuations: without them, we would not be here to discuss them.

Structures grow

The Cosmological Principle holds that, on the grand scale, the universe is both homogeneous (the same, on average, in all parts) and isotropic (the same, on average, in all directions) – facts confirmed both by observations of the recession of the galaxies and measurements of the temperature of the cosmic microwave background, identical in all directions. The concept of inflation provides an explanation for the observed homogeneity and isotropy. However, there is nothing to prevent there being small-scale irregularities in the fabric of the cosmos. Indeed, this is what we observe in the case of galaxies, stars and planets. This is an essential point: if there had been no irregularities in the primordial universe, it would have remained totally homogeneous, and no condensed object could possibly have formed. We therefore must imagine the presence of 'clumps' in the primordial cosmic fluid. Even though they were very small, these overdensities created local gravitational fields. A clump would tend to grow, since the overdensity attracts material inwards from its environment. *A priori*, this growth would have continued ceaselessly, since nothing can halt the action of gravitation. However, as we have seen, photons were trapped by baryons until the time of the recombination, and remained with them. They shared a common destiny. Only radiation pressure tended to oppose the growth of clumps.

Celestial harmonies

These opposing actions have definite consequences which we can illustrate by considering the well known case of sound waves. If we compress air, for example by banging the membrane of a drum with a stick, we create a sound wave which propagates through space with time. The small local increase in air pressure which has been created will lessen as the gas resists the compression, but there is no immediate return of this element of the gas to its initial state, and the compression moves, forming another overdensity a little further along, and so forth: we have created a sound. The phenomenon is very similar in the primordial universe: the battle between gravitation and the resistance to pressure created a succession of overdensities and depressions in the fluid. A local perturbation produced oscillations which propagated through the cosmic fluid like sound waves though air. We therefore call these cosmic acoustic waves. These cosmic waves manifest themselves as a series of troughs and peaks of density, causing corresponding variations in temperature (when a gas is compressed, it is heated, as we experience when using a bicycle pump). As these successive contractions and expansions occur, the temperature of the fluid, which is directly related to pressure, undergoes positive or negative variations.

When the recombination happened, photons were no longer held captive and were finally able to escape from other photons freely, preserving the 'memory' of the (slightly perturbed) temperature of the regions from which they came. An essential prediction of the cosmological model is therefore that we should observe tiny differences in temperature in different zones of the sky. The amplitude of these variations will reflect the size of the initial clumps, and this is indeed what we observe. The most recent measurements made by the WMAP team suggest an amplitude of the order of one part in 100,000 (in agreement with the theoretical predictions) for the variations in temperature.

However, it is possible to analyze further these miniscule fluctuations. A detailed analysis of the phenomenon of cosmic waves reveals the existence of a particular spatial scale. In effect these cosmic waves have only a limited propagation time, the time between the creation and the epoch of recombination, which limits the distance for which they can propagate. We infer from this a characteristic scale: the acoustic horizon. The size of this horizon can be estimated, considering that the velocity of sound in the cosmic fluid is of the same order as that of the speed of light, assuming the duration to be the time between the Big Bang and the recombination. Using these values, we obtain a dimension of the order of 450 million light years. At the instant of recombination, photons were finally free from baryons and were able to leave these small overdensities behind. Even if the actual situation is more complex than the one just described, the existence of an acoustic horizon and therefore of a privileged scale remains. We await the discovery therefore, by studying the distribution of temperatures of the cosmic microwave background in the sky, of the particular spatial scale corresponding to this acoustic horizon. In reality the clumps at the origin of temperature fluctuations were scattered randomly within the cosmic fluid and this caused a certain 'cacophony' in the celestial music. It is therefore necessary to use statistical techniques in order to extract the required information.

Without entering into too much technical detail, we should mention the deployment of a tool known as the 'power spectrum' for measurements of temperature. This is a mathematical quantity enabling us to investigate whether, on a given scale, identical characteristics exist everywhere on the chart examined. If this is not the case, then this quantity is zero for that particular scale. Then the calculation is resumed on a different scale. In the opposite case its value will differ from zero. If such a statistical tool is used, the acoustic horizon will manifest itself through a significant signal at the predicted spatial scale. In practice, angles rather than distances are involved in the analysis of sky maps. Converting the predicted value of 450 million light years to angle è, we obtain, for the acoustic horizon, a value of è $\sim 1^\circ$. To

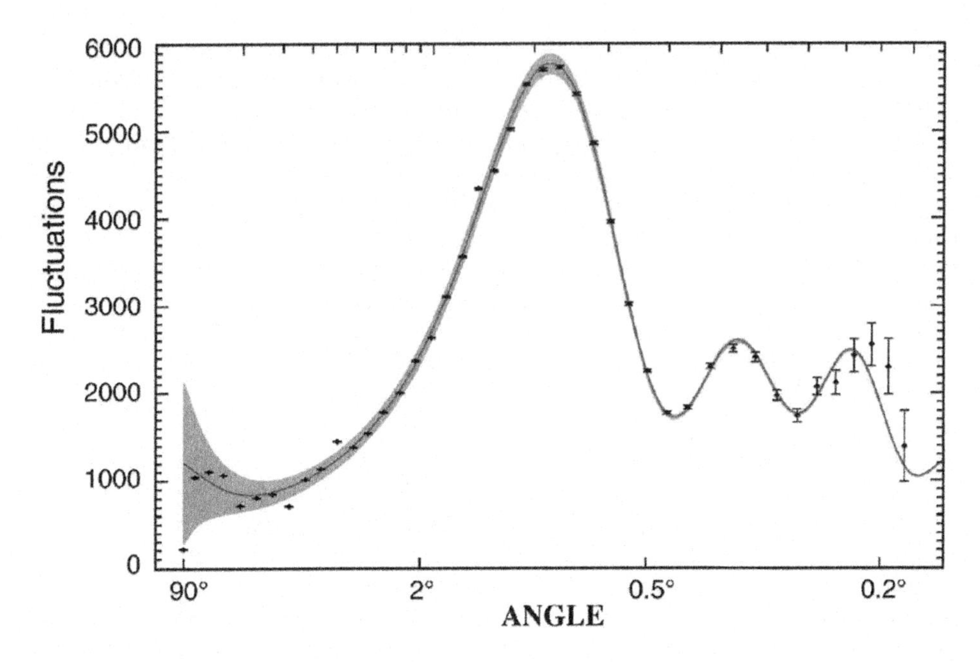

Figure 5.4 The angular power spectrum of small fluctuations in temperature as measured by the WMAP spacecraft. It shows the relative brightness of the 'spots' in the WMAP full-sky map versus the size of the spots. This tool enables us to identify a 'privileged distance' through the presence of one or more peaks at particular scales. When studying the cosmic microwave background, we use angles on the sky shown here on the horizontal axis, decreasing from left to right. Note the presence of a very pronounced peak at an angle of approximately 1°, corresponding to the existence of a characteristic size: that of the acoustic horizon (450 million light years) (NASA/ WMAP Science Team).

the great satisfaction of cosmologists, this is precisely the angle observed at the pronounced peak of the signal in the power spectrum of temperature as measured by the WMAP mission (Figure 5.4).

This has been an immense success for both theorists and experimenters, the fruit of decades of effort, confirming the cosmological model in its description of the universe, from the Planck era to a time almost fourteen billion years later. (It is also a measurement of the quantity of baryons at the time of the recombination, from the height of the secondary peaks, 'harmonics' of the first: see below).

What more could we ask for - prediction and observation in accord! Note that, in Figure 5.4, the WMAP data produce not just one peak but a series of peaks and troughs whose intensity decreases with the scale considered. This is due to the fact that, at the moment of recombination, the waves could have been in different states, either condensed or rarefied. Indeed, complete and detailed calculations predict such harmonics, like notes from a musical instrument. Let us underline again the perfect accord between the theoretical predictions and the observations. The detection and measurement of these acoustic peaks, resulting from the growth of primordial perturbations of quantum origin engendered during the inflationary phase of the universe, is certainly one of the greatest triumphs of the modern cosmological model!

Weighing matter and light

Now let us leave the photons of the cosmic microwave background for a moment, and resume our quest for baryons. We in fact completely neglected them in the course of our estimates above of the speed of sound through the cosmic fluid. Although they are *a priori* quite negligible (there being a billion photons for each baryon), the part they play ought to be detectable, since their presence modifies the speed of the sound involved in the propagation of waves. In fact, this speed is a function of the density of the baryons; the greater the density, the slower the speed of sound. This means that the acoustic horizon will become smaller as the fraction of baryonic matter increases. Moreover, in the competition between gravity and pressure, baryons add their mass to gravity, bringing about a compression entailing greater variation in temperature. The amplitude of the peaks and troughs in the power spectrum of temperature will therefore depend on the quantity of baryons in the cosmos at the time of the recombination. Measurement of the peaks leads to the deduction of the value of Ω_b at an epoch when the universe was only 300,000 years old.

Measuring these (infinitesimal) temperature fluctuations for the whole sky and extracting scientific information about, for example, the density of baryons is no small affair. In recent decades, many experiments have been carried out, both ground-based and in space. The first of these was COBE (Figure 5.5), a satellite built under the aegis of NASA and launched in 1989. This led to a Nobel Prize for the COBE team's principals in 2006, a recognition of the first measurements to achieve consensus among the scientific community. Its successor, WMAP (Figure 5.6), launched by NASA in 2001, achieved unrivalled accuracy after five years of continuous measurements, a consummation of earlier estimates carried out by ground-based instruments and balloons (see chapter 8). In the final analysis, the value for the baryonic density (present epoch) was settled, after WMAP, at

$$\Omega_b = 4.4 \text{ percent}$$

Let us take a moment to note that this value was obtained totally independently of that derived from the study of primordial nucleosynthesis,

Figure 5.5 Artist's impression of the COBE spacecraft which was the predecessor to WMAP. COBE was launched by NASA into an Earth orbit in 1989 to make a full sky map of the cosmic microwave background (CMB) radiation leftover from the Big Bang. The first results were released in 1992. COBE's limited resolution (7 degree wide beam) provided the first tantalizing details in a full sky image of the CMB (NASA/COBE Science Team).

and the values are remarkably similar: here is one of the strengths of the Big Bang cosmological model. Finally, let us point out that the different categories of energy/matter intervene in intricate ways in the physics of recombination. This complexity is also an advantage, because apart from Ω_b, temperature measurements of the cosmic microwave background allow us to measure the total energy/matter content, characterized by the parameter Ω_T (given by the position of the first peak) and the total matter content Ω_m (both for the present epoch). The WMAP analysis leads to the remarkable result (Figure 5.7):

Figure 5.6 The WMAP spacecraft used the Moon to gain velocity for a slingshot to the L2 Lagrange point. After three phasing loops around the Earth, WMAP flew just behind the orbit of the Moon, three weeks after launch. Using the Moon's gravity, WMAP steals an infinitesimal amount of the Moon's energy to maneuver out to its final position at the L2 Lagrange point, 1.5 million kilometers beyond the Earth (NASA/WMAP Science Team).

$$\Omega_T = 100 \text{ percent} \quad \Omega_b = 4.4 \text{ percent} \quad \Omega_m = 23 \text{ percent}$$

confirming that, on the one hand, baryonic matter is not the dominant component of matter in the universe, and that, on the other hand, there exists dark energy of density Ω_X (of the order of 72 percent, since the sum of the components must be equal to Ω_T = 100 percent).

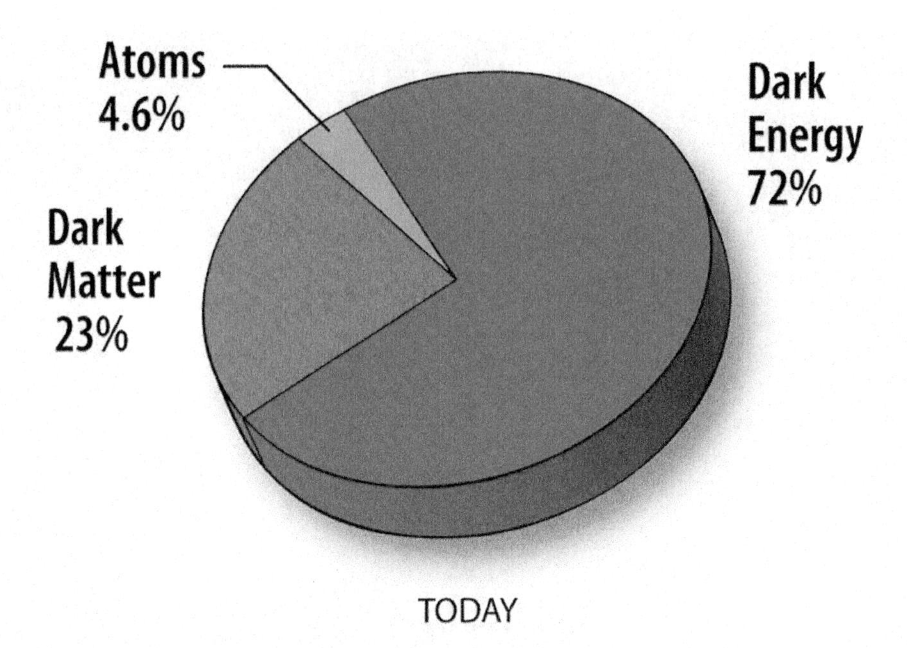

Figure 5.7 Astronomers' estimates, from WMAP data, of the energy/matter content of the universe today (z = 0) show that it consists of 4.6 percent atoms, 23 percent dark matter, 72 percent dark energy and less than 1 percent neutrinos. So, ordinary ('baryonic') matter, essentially responsible for the radiation emitted by stars and galaxies, represents but a small fraction ($\sim$5%) of the global contents of the cosmos, which is dominated by dark matter and dark energy. Dark matter does not emit or absorb light; it has only been detected indirectly by its gravity. Dark energy acts as a sort of an 'anti-gravity'. This energy, distinct from dark matter, is thought to be responsible for the present-day acceleration of the universal expansion (NASA/WMAP Science Team).

6 Cosmic canvas

"Blunt scissors spoil good cloth." Guadelupian proverb

We have sought out and weighed the baryons in the nearby universe, and to our great frustration have had to resign ourselves to the fact that we did not find the entirety of our legacy from the primordial nucleosynthesis three minutes after the Big Bang. However, celestial harmonies in the cosmic microwave background from 300,000 years later proclaimed that 'all was gathered in'. So what has happened between then and now, 13.7 billion years later? What became of the baryons in all that intervening time? These are yet more turnings along our pathway, and they will have to be investigated.

Structures rule

After the capture of electrons by protons and the nuclei of heavier elements, rendering the cosmos neutral and leading to the formation of the cosmic microwave background, the universe was composed only of dark matter and neutral gas (75 percent hydrogen, 25 percent helium, and traces of deuterium, lithium and beryllium from the primordial nucleosynthesis). This gas was perfectly chemically inert in the conditions then prevailing in the universe (temperatures of thousands, and later hundreds of degrees). Apart from the expansion of the universe, absolutely nothing happened for millions of years... *Absolutely* nothing?

Well, not quite, as gravity, until then only a minor player, began to find more favorable conditions for its workings. Now, the growth in fluctuations of matter could not really get under way until matter at last dominated radiation. Gradually, the clumps (born of primordial quantum fluctuations, as described in the previous chapter) could begin to grow under the effect of their own mass: very slowly at first, for they were scarcely denser than the gas surrounding them (initially, the difference in density of the dark matter was but one part in 100,000). Then, as their density increased, these clumps, composed essentially of dark matter, attracted more and more material around them, thereby adding to their gravitational potential. They would finally become 'dark halos', in equilibrium, the progenitors of future galaxies, within which baryons (in the minority) would also be coming together to form the stars that we see. Within these concentrations, the gas eventually

Figure 6.1 A representation of the evolution of the universe over 13.7 billion years. The far left depicts the earliest moment we can now probe, when a period of 'inflation' produced a burst of exponential growth in the universe. (Size is depicted by the vertical extent of the grid in this graphic.) For the next several billion years, the expansion of the universe gradually slowed down as the matter in the universe pulled on itself via gravity. More recently, the expansion has begun to speed up again as the repulsive effects of dark energy have come to dominate the expansion of the universe. The afterglow light seen by WMAP was emitted about 300,000 years after inflation and has traversed the universe largely unimpeded since then. The conditions of earlier times are imprinted on this light; it also forms a backlight for later developments of the universe (NASA/WMAP Science Team).

attained the kind of temperatures and densities that could trigger nuclear reactions. So, after some tens of millions of years of chemical inertia (the so called Dark Ages, see Figure 6.1), suffused only by dull fossil radiation, the universe came alight, here and there, with the first stars...

In this way, little by little, the grand structure of the universe was built up, within which today's galaxies are organized in a network of filaments and sheets around great voids, with, at their intersections, clusters of galaxies, the knots binding the very fabric of the cosmic canvas.

The first stars and quasars

The first stars to shine in the cosmos lit up at least ten billion years ago. Were they very different from our star, the Sun? This is thought to be almost a certainty, since the Sun is a relative late-comer, only five billion years old. It was formed in an environment already enriched with elements synthesized in earlier-generation stars. As will be seen, this presence or absence of heavy elements is a key factor in the genesis of stars. It is of course very difficult to investigate *via* observation the nature of first-generation stars (their mass, luminosity, lifetime, radiation, etc.).

Astronomers, true to their reputation for illogicality in nomenclature and numbering, call this primordial population of stars - Population III stars. They were short-lived, as we shall see, and the primordial environment where they formed no longer exists today, or has at least changed in its nature. Recently, the discovery of very remote gamma-ray bursts (at redshifts $z > 8$) has revived hopes that we may be able directly to detect the deaths of some of these primordial stars in colossal explosions. Sadly, it seems likely that we have only a partial view of these objects, assuming that we are able to detect them at even greater distances. Alas, the most remote objects yet observed are representative only of epochs later than that of the formation of the first objects. Given such paucity of data, what can we really find out about this population?

An essential factor in the process of stellar formation is the chemical composition of the medium within which the stars are created. This composition is a fundamental ingredient in the cosmic recipe that determines the initial mass of stars. In effect, the presence or absence of metals (in astrophysics, elements other than hydrogen and helium) regulates the radiative cooling of gas, thereby determining its temperature, pressure and ability to resist gravitational collapse and fragmentation. How is this played out? When a gas contracts under the effect of its own gravity, its pressure and density increase. Collisions between the atoms of the gas become more frequent, and some electrons pass to higher energy levels. When these electrons return to their initial levels, the atom emits a photon whose energy corresponds to the difference between the two energy levels, initial and final. The photon, having been emitted, will escape from the cloud because the density of the gas at the beginning of the process is not sufficient to support interactions between electrons and photons, therefore depriving the contracting gas of some of its energy. This energy loss causes what is known as 'radiative cooling' of the cloud. This phenomenon is reinforced by the presence of atoms other than hydrogen and helium, because that presence increases the number of possible atomic transitions by electrons.

The number of collisions favoring energy transfer also increases. A collision producing less energy than the smallest difference in energy accessible within a hydrogen atom in its fundamental state would therefore not produce a photon. The presence of additional atoms of different kinds with their different energy levels means that more collisions can occur. A gas containing metals is therefore more efficiently cooled than a hydrogen/helium gas during contraction. Obviously, this cooling will cease as soon as the medium becomes very concentrated and photons can no longer escape with their energy, as they are now trapped by their interactions with matter. The situation within this cloud is analogous to that existing for the universe as a whole before the recombination. This allows the gas therefore to reach temperatures high enough for nuclear reactions to begin.

Having examined one of the factors essential for stellar formation, we can now return to the genesis of the very first stars. In reality, a cloud of primordial gas is neither perfectly spherical nor perfectly homogeneous. Physical conditions, principally those of temperature and pressure, vary from location to location, and the consequence of these variations is that, as contraction proceeds, the cloud divides into several fragments, each one the birthplace of a star. The size of these fragments depends upon 'local' physical conditions. Calculations show that, the lower the temperature for a given density, the more fragments there will be. So, in a cloud of primordial gas (by definition lacking in metals and therefore cooling only slowly), the (relatively large) fragments may give rise to quite massive stars, up to 300 times the mass of the Sun. In environments rich in heavy elements, however, the upper limit will be more like 60-80 solar masses.

There are two direct consequences of the fact that Population III stars have higher masses. The first is that their lifetimes are very short (sometimes less than a million years), and they are destined to end their days in a supernova explosion (Figure 6.2): this explains why we find no trace of these stars today, but we might be able to detect their deaths by observing the most distant gamma-ray bursts. The second consequence is that they will have produced extremely intense ultraviolet radiation, and we shall examine the importance of this later.

We have been discussing the earliest Population III stars, but let us not forget that the galaxies containing them were formed in concomitant fashion! In the range of overdensities there were those concentrated enough to become the birthplaces of the first black holes, objects so massive that light itself cannot escape from them. Once formed, these cosmic ogres would *accrete* matter from their surroundings in the parent cloud, and this matter would gradually form into a disk (accretion disk) rotating around a central body. Particles accelerated within this rapidly rotating disk would emit a specific kind of radiation known as *Bremsstrahlung* or free-free emission. The spectrum of this radiation, very different from that of stars, is the characteristic signal of a new type of object: quasars (quasi-stellar objects, referring to the fact that they appear as point sources in the sky). This spectrum, which reflects the evolution of the luminosity as a function of wavelength, is characterized by its relatively flat nature, unlike the more 'humped' form seen in that of stars (Figure 6.3).

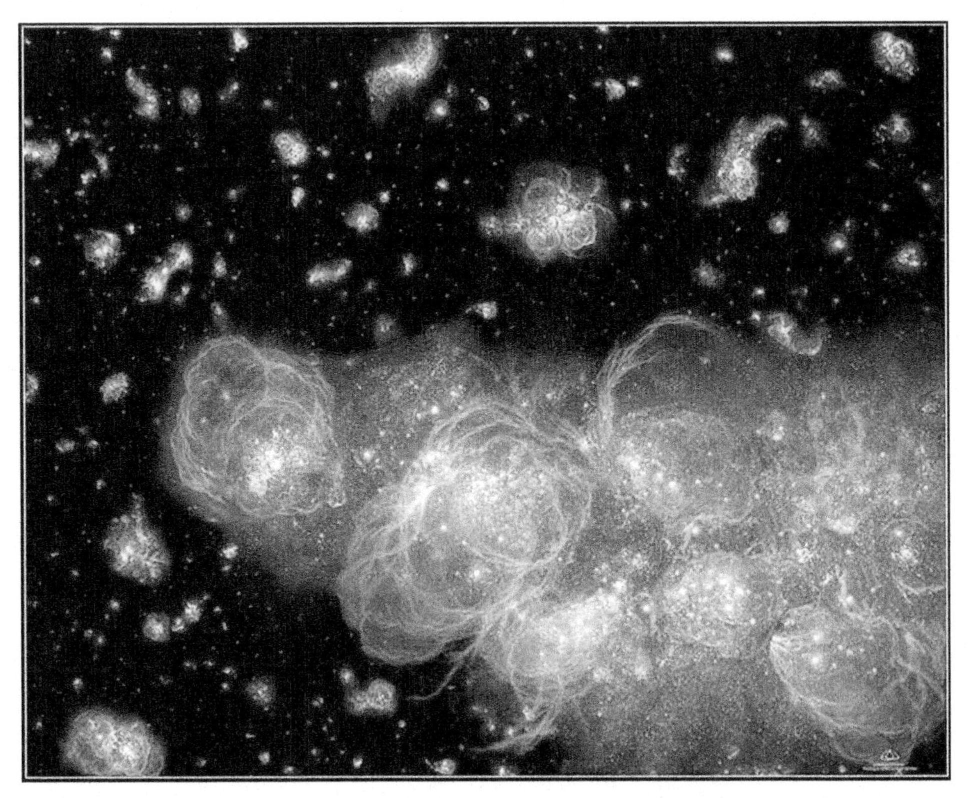

Figure 6.2 An artist's impression of how the very early universe (less than one billion years old) might have looked when it went through the voracious onset of star formation. The most massive of these Population III stars self-detonated as supernovae, which exploded across the sky like a string of firecrackers (Adolf Schaller for STScI).

Most of a quasar's energy is emitted in the form of ultraviolet and X-rays, very energetic domains of the electromagnetic spectrum: the photons of these radiations are easily able to strip electrons from atoms during interactions with matter. As will be seen, the quasars, amazing objects in their own right, will play a major role in the pages to come.

Cooking with leftovers

So baryons were destined to become concentrated within halos of dark matter to form the first stars. One question comes immediately to mind: was all the initial supply of gas transformed into stars? In other words, was the process 100 percent efficient? If the answer is yes, 'weighing' the baryons of these remote epochs is no easy task, since the stars made then

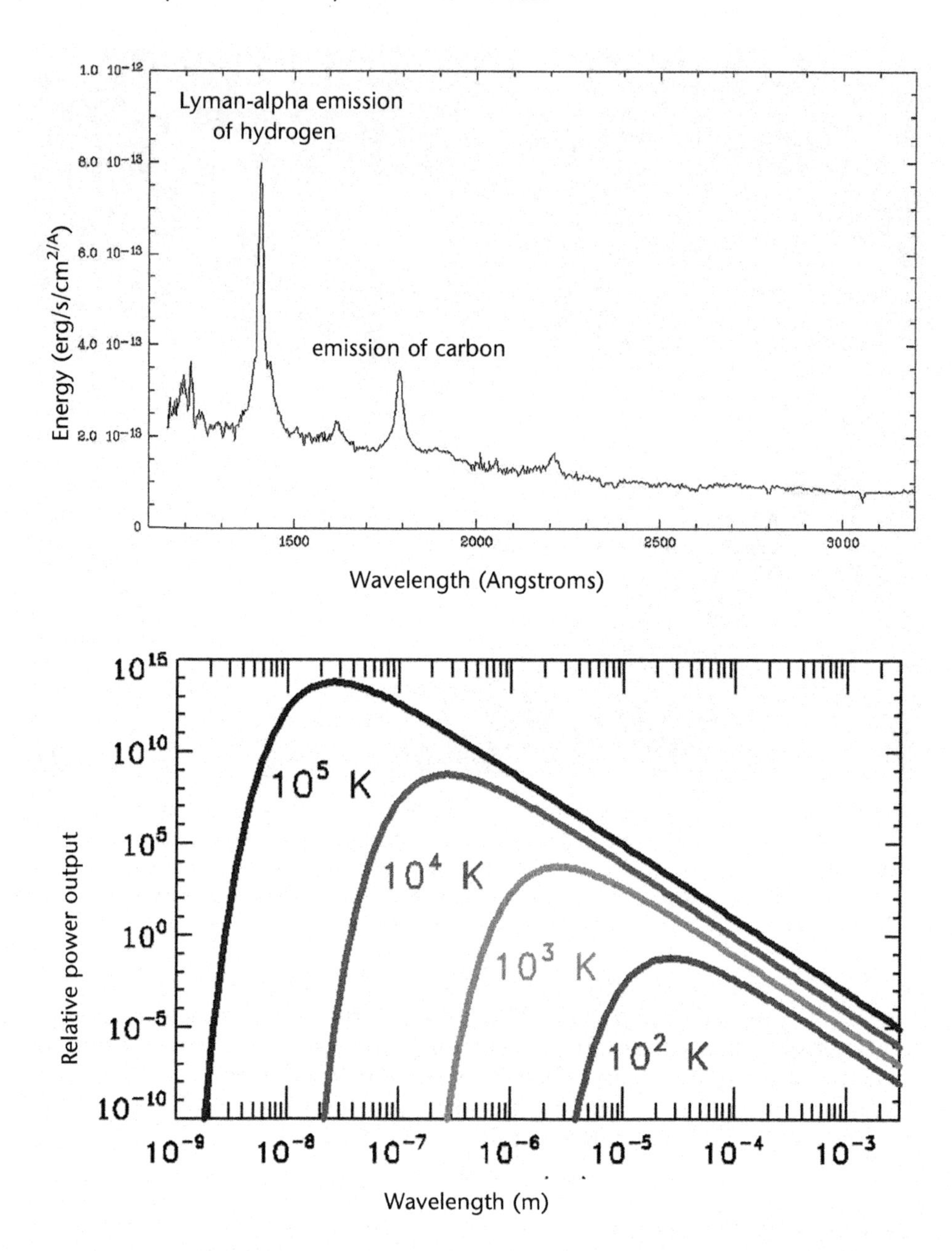

Figure 6.3 Comparison of the spectra of stars and quasars. The spectrum of the nearby quasar 3C273 (top), the brightest in the sky, showing the characteristic Lyman-alpha line, and spectra for a black body at different temperatures (bottom). Broadly speaking, the spectra of normal stars closely resemble that of a black body. Note that the ultraviolet flux is increased in the case of the quasar, whilst that of the Sun is insignificant compared with its visible flux.

Fact box More about quasars

The first quasars were discovered in the early 1960s during a program aimed at the identification of radio sources detected by the Cambridge interferometer. One of these apparently stellar sources exhibited a strange spectrum, comprising very wide lines, corresponding to no known element. At first, astronomers thought they were dealing with a new kind of star. Other sources with identical characteristics were then found, but their nature remained a mystery – until Maarten Schmidt, an astronomer at Caltech, had the idea of applying a redshift of 0.15, ten times greater than any value hitherto measured, and identified the known lines of hydrogen and oxygen.

At this distance the measured luminosity implied that these sources were intrinsically very bright indeed, of the order of several tens of times brighter than a normal galaxy, and concentrated within a very small region. Only later was the hypothesis proposed that these strange objects were massive black holes surrounded by disks of matter rotating at high velocity. This hypothesis also explained their compactness, energy output and the nature of their spectra.

The lines observed in the spectrum of a quasar are wide, because the gas is very agitated (we can well imagine that the environment of a black hole is not the calmest of places). An atom which emits a photon of wavelength λ_0 when it is at rest will emit at wavelength $\lambda_0(1 + v/c)$ when its velocity is v (and c is the speed of light). So a collection of atoms which have velocities between, for example, $-$ v and + v (atoms may be approaching as well as receding) will emit at all wavelengths between $\lambda_0(1 - v/c)$ and $\lambda_0(1 + v/c)$. In the regions surrounding quasars the velocity of the gas may be greater than 5,000 kilometers per second (compare this with the typical velocities of, at most, 300 kilometers per second found in gas clouds of 'normal' galaxies); it is for this reason that the lines are very wide.

have by now disappeared. If the answer is no, where is the unused gas? What fraction of the original amount does it represent? Can this gas be detected, if it indeed exists? We shall be investigating these questions - but first, a little detour...

We have seen that the primordial stars and quasars had the particular characteristic of emitting very energetic ultraviolet photons. Such radiation may well have re-ionized the rest of the gas not made into stars. If gas were present, this phenomenon would manifest itself first in bubbles (known as Strömgren spheres) surrounding the earliest objects formed; and then throughout space, as the number of stars and quasars gradually increased, and as more and more energetic photons were emitted.

It will be immediately seen that any left-over gas, whether it represents a mere fraction or, on the contrary, most of the initial baryons, could be in some neutral or ionized form depending on the epoch (i.e. on the redshift) at which it is observed. We can easily imagine that the residual gas is very tenuous, since if it were not, it would have become 'star-stuff', and that it is spread out along the filamentary structure of galaxies, constituting what is known as the intergalactic medium (IGM). We also know that gas composed of hydrogen (not forgetting the helium) in these conditions of density and temperature cannot produce emissions detectable by our current instruments. It may well be that the instruments of the future, such as the Square Kilometre Array (SKA), will be able to detect such emissions (see chapter 8). So how can we reveal their presence?

To the great good fortune of the astronomers, gas can also absorb radiation (the Earth's atmosphere, which absorbs the Sun's ultraviolet radiation, is the best example, protecting life on our planet). Intergalactic gas is no exception to this rule. In order to absorb a photon, an atomic nucleus should be accompanied not only by at least one electron (as is the case with neutral hydrogen and even singly ionized helium), but also be in the path of an exterior source of photons. The 'fossil' cosmological radiation, although omnipresent, cannot perform this function. Indeed, atomic hydrogen absorbs principally in the ultraviolet, while fossil radiation, even if the universe was a little hotter at the time of the earliest objects, emits mainly in the radio domain. Fortunately, quasars, or QSOs (quasi-stellar objects) as they are sometimes known, are there and come to the rescue.

They have the triple advantage of being compact (offering a single line of sight), highly luminous (easy to observe even at very great distances), and emitting most of their radiation in the ultraviolet. This is why these objects are used as cosmic beacons lighting our way through the intergalactic medium (Figure 6.4). How can the possible presence of gas along the line of sight to a remote quasar manifest itself? We can gain some idea of this by looking again at the spectrum of the quasar 3C273 (Figure 6.3a). We see, as a function of wavelength, a flat form (the *Bremsstrahlung* mentioned above), and what is essentially a very wide emission line: that of Lyman-alpha.[1] These wide lines (we also see another, Lyman-beta, which is much less intense) correspond to emissions from regions on the outside of the

---

[1] The Lyman series are a sequence of spectral lines (Lyman-alpha, beta etc.) emitted by hydrogen. A hydrogen nucleus captures a free electron at the first level of its electrical potential, and the surplus energy is liberated in the form of a photon at a wavelength of 1215 Å (Lyman-alpha); the second excitation level corresponds to the Lyman-beta line, and so on... These lines also appear as absorption features when the gas absorbs photons at appropriate frequencies.

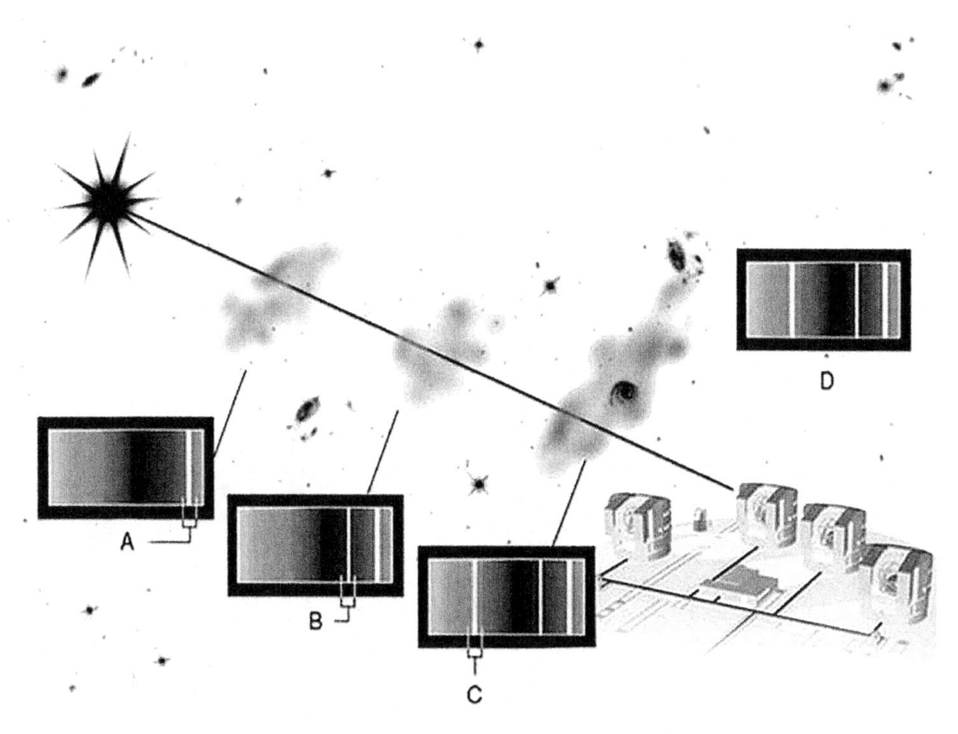

Figure 6.4 Quasars as cosmic markers. The line of sight to a quasar may pass through several gas clouds which leave their signatures in the observed spectrum at different wavelengths, depending on the distance and therefore the velocity of the clouds (ESO).

accretion disk, zones close to the black hole, highly excited by its radiation and re-emitting the photons absorbed at the characteristic wavelengths of the elements composing them. These spectacular hydrogen emission lines are, however, not the spectral characteristics that interest us here, as they are linked to the immediate vicinity of the quasar, and not to the intergalactic medium that we are trying to reveal.

A mysterious forest

Let us now consider the case of a quasar at redshift $z = 2.5$, the distance corresponding to a time when the universe was 2.5 billion years old (Figure 6.5). We find the same Lyman-alpha line here, but redshifted towards longer wavelengths as a result of the expansion of the universe, its presence characteristic of hydrogen.

However, what is striking when viewing this spectrum is the asymmetry

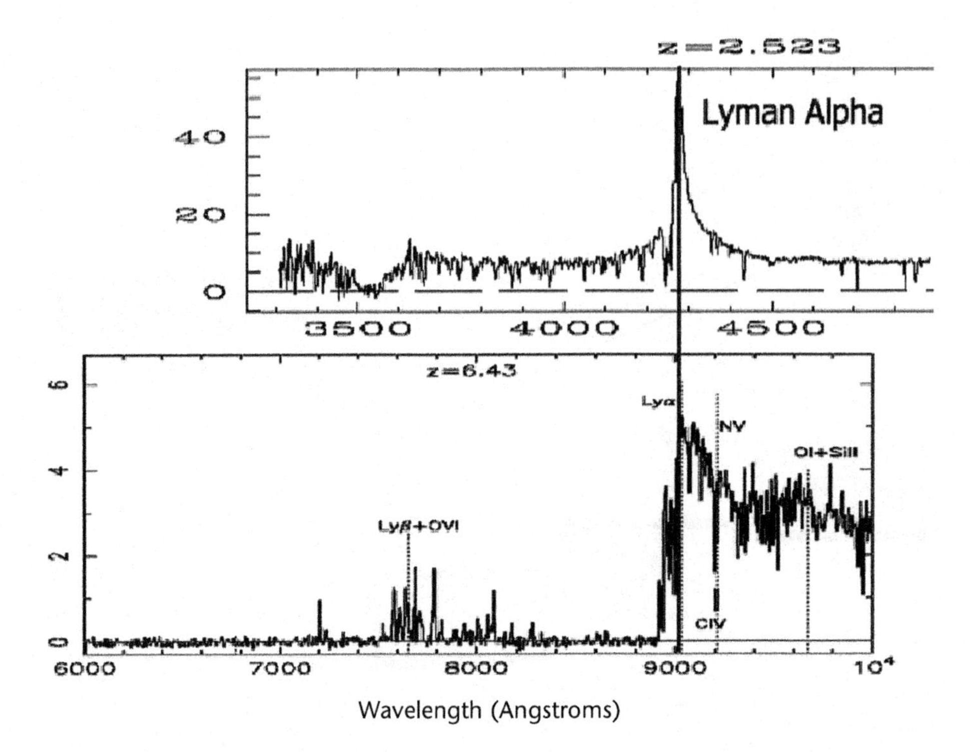

Figure 6.5 (Top) A quasar at redshift z = 2.5. (Bottom) A quasar at redshift z = 6.43, one of the most remote. If we compare the spectrum of the quasar at redshift z = 2.5 with that of 3C273 (see Figure 6.3), we notice the appearance of absorptions (dips) due to intergalactic clouds. The spectrum of the quasar at redshift z = 6.4, observed at an epoch just after the re-ionization of the universe, shows total absorption to the left of the Lyman-alpha peak, with a few emission lines (Dr F. Briggs).

between the section to the left of the Lyman-alpha emission line, and that to its right. The right-hand section shows only occasional, isolated absorption lines, corresponding to heavy elements present in a single galaxy lying in the line of sight to the quasar, but the left-hand section is remarkable for its very large numbers of absorption features. This difference can only be explained by inferring the presence of neutral intergalactic gas clouds along the line of sight, each one responsible for a Lyman-alpha absorption line. The fact that these lines are separated on the spectrum (normally they correspond to absorption at 1215 Å) is merely a function of the expansion of the universe. Since the velocity of recession of a galaxy or a cloud is proportional to its distance (according to Hubble's law), two clouds lying at different distances will absorb radiation at different wavelengths. In the present case, the universe makes life easier for us by separating the signatures of absorption by different clouds (which, if they were on the same line of sight and motionless relative to each other, would show superposed absorption and be indistinguishable).

Let us now turn to the most distant quasars yet detected ($z \sim 6.5$, Figure 6.5). We see that the Lyman-alpha emission line lies at a relatively long wavelength as a result of the expansion of the universe (cosmological redshift), while the spectrum is totally absorbed to its left. This must likewise be due to the presence of a large number of intergalactic clouds, or even to an as yet totally neutral medium along the line of sight. The result in either case is the total absorption of the flux of the quasar. So we see that this Lyman-alpha forest, as the astrophysicists who discovered it call it, reveals to us the presence of large amounts of neutral gas either as clouds or in the form of a continuous medium. In the latter case, the transition between total absorption and absorption in the form of separated lines will indicate the redshift, and therefore the date of re-ionization. In-depth analysis of the data obtained from high-redshift quasars shows that this transition probably occurred at $z > 6$, while independent measurements by WMAP place it at $z \sim 17$.

Further investigations will be required to achieve better compatibility in the results. However, not only can we determine dates and distances, but we can also find out more, especially about the quantity of gas causing the absorption. By extrapolation, taking into account the intensity of the radiation, and using calculations from atomic physics which allow us to deduce the screening capacity of the gas, we can work out the total mass of these clouds. It must however be remembered that measurements of absorption in the spectrum of a single quasar give information only on the 'column density' of the gas, i.e. the number of atoms projected onto a very small area of the sky along the line of sight. To arrive at a significant figure, we would have to determine the spatial extent of the absorbing zones.

At first sight, this is a simple procedure: with enough lines of sight through the same clouds, we can measure their sizes in the sky. Quasar hunters also have a heaven-sent advantage: there exist groups of quasars, separated by just seconds or minutes of arc, representing actual separations of tens to hundreds of thousands of light years, similar to the size of intergalactic clouds. However, these groups are rare, and we cannot choose separations to test the different sizes. The fact that we have detected a cloud situated on two different lines of sight gives only a minimum value for its size. We fall back upon statistical estimates, comparing the absorptions present on one or more lines of sight and deducing the mean distribution of the sizes of the clouds. After all this painstaking work, made possible only by the advent of high-performance spectrographs on very large telescopes (so remote and faint are the celestial targets) such as the VLT at the European Southern Observatory or the Keck Observatory in Hawaii, we can proceed to the end of the exercise. The most recent results show that, taking into account the uncertainties inherent in the method, the Lyman-alpha forest contains more than 90 percent of the baryons of the universe for redshifts $z > 3$. Add to that the 3 percent of baryons in galaxies that had already been formed at that epoch (to which correspond the so-called 'metallic' lines, for example the CIV carbon line to the right of the Lyman-alpha line), and we arrive at something like the correct total...

At z = 3, as at z = 1000, it works out.

The final reckoning

There is nothing to prevent us using this same method in the case of the nearby universe ($0 < z < 3$) to try to estimate the quantity of baryons we might find in the 'forest' and therefore not in the stars, galaxies and clusters of galaxies mentioned in previous chapters, and whose contribution, remember, was only 0.5 percent or ~1/8 of the primordial baryons. It is obvious from the spectrum of the nearby quasar 3C273 that the 'local' forest is much more thinly populated than that of the earliest epochs of the universe. Not at all an unexpected result: in the course of the last 13 billion years, the universe has seen the incessant creation of stars and galaxies, drawing on its reservoir of primordial gas, and gradually exhausting the supply. Moreover, physical conditions within the intergalactic medium have changed, mostly because of the lessening density of quasars. So what is the quantity contained within this local forest? We might at first hope to discover the missing 7/8, i.e. if the contribution of the local Lyman-alpha forest were 3.5 percent. Alas! Once again, we fall short, since we arrive at only 1.8 percent and therefore a total of 2.3 percent for all the contributions at z = 0, according to our previous estimate.

This means that about 50 percent of the baryons of the primordial nucleosynthesis **are still missing** (Table 6.1 and Figure 6.6). Indeed, the nearby universe is still playing nasty tricks on us! The missing 50 percent still awaits discovery; new tools involving computerized simulations will be needed for this task, as we shall see in the next chapter.

Table 6.1 Fractions of ordinary (or baryonic) matter existing as stars and gas within galaxies, hot X-ray gas in clusters and in the local Lyman forest, compared with total energy/matter in the universe, expressed as different parameters of Ω_b.

Nature of ordinary matter	Contribution to total energy/matter content	
Stars + gas in galaxies + hot X-ray gas in clusters	$\Omega_{b\text{-stars/gas}}$	~0.5 percent
Local Lyman forest	$\Omega_{b\text{-Local-Lyman-alpha}}$	~1.8 percent
Total (z = 0)		**~2.3 percent**

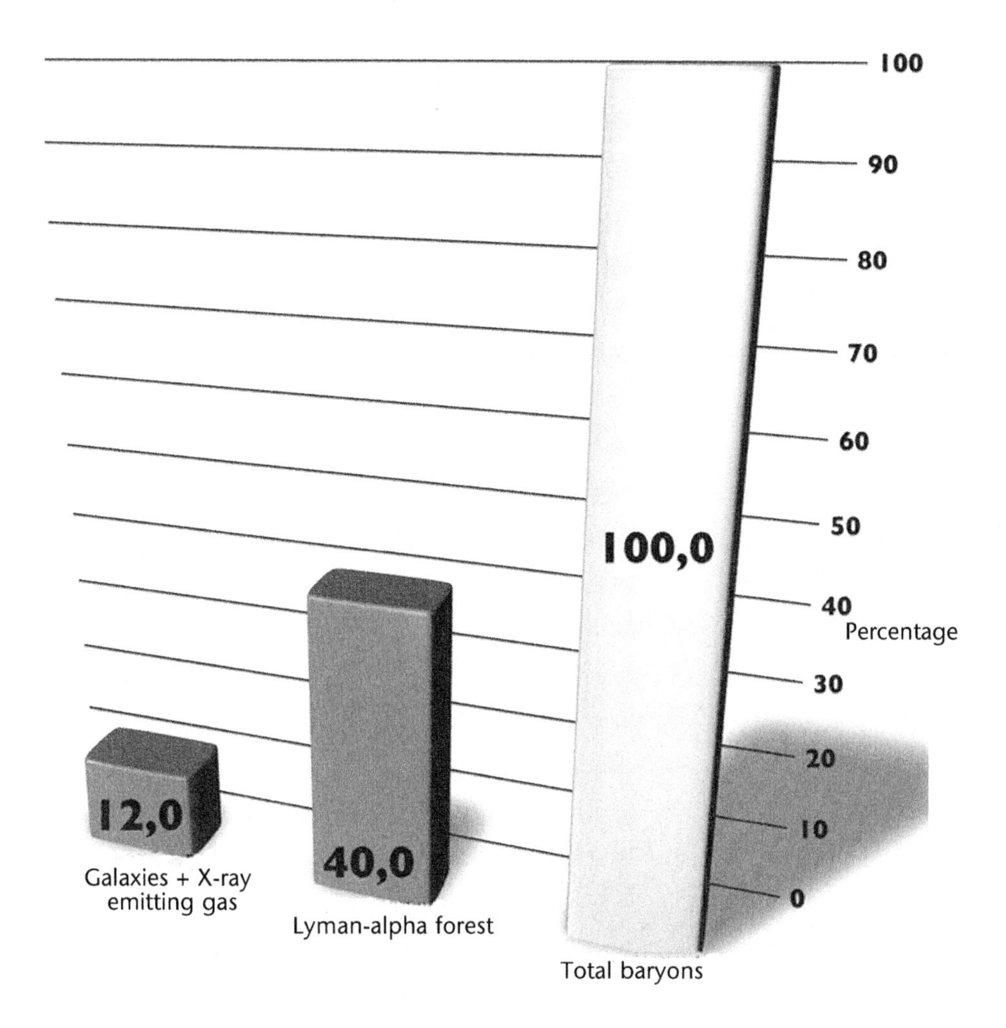

Figure 6.6 Comparison of quantities of ordinary matter detected as stars and gas within galaxies, hot X-ray gas in clusters and in the local Lyman-alpha forest, with the total contents of this baryonic matter in the cosmos. Nearly 50 percent of ordinary matter is still unaccounted for.

7 Lifting the veil: simulations

"He who knows not how to dissimulate knows not how to reign." Louis XI

As we have seen in the preceding chapters, a significant fraction (about one-half) of the baryons in the universe have still to be accounted for at present, while the quantity of dark matter, whose nature remains an enigma, seems to have been fairly well calculated. A fine state of affairs! So how and where have they hidden themselves? In the previous chapter we sketched out scenarios for their evolution which seemed convincing, at least for redshifts of the order of $z = 3$. For more recent epochs (or smaller z), this ordinary matter, although still relatively local, escapes us; it seems to be beyond our capacity to observe it. So we have to fall back upon simulations to delve into this mystery.

Why simulations? Unlike their colleagues in physics and chemistry, astrophysicists, and in particular cosmologists, cannot perform their experiments in laboratories in order to study the objects of their researches under different conditions in an attempt to analyze and define their properties. They cannot, unlike their colleagues the planetologists, bring back samples for study. The objects that interest astrophysicists are simply unreachable, and it is not feasible to create a star or a galaxy in a laboratory. There is the exception of complex molecules found in the interstellar medium, which researchers can recreate and study on their benches, although the conditions of vacuum and temperature of that environment cannot be duplicated by our current equipment. The study of galaxies and stars in their various environments may well be envisaged, in an attempt to deduce their broad properties (*via* statistical analyses), but the observable universe itself, as cosmologists see it, is unique, even though some evoke the hypothesis of the multiverse. Certain models envisage not just one single universe but the possible existence of a very large number of universes (for example, $\sim 10^{500}$!), within each of which the constants of physics could be different from ours, and which would therefore all have different destinies.

Two difficulties arise from this, because we have only one *realization*[1] of the

---

[1] If a process leading to the establishment of a value or set of values is random, i.e. dictated by the laws of chance, this value or set of values is called the *realization* of the process. This is the case, for example, when tossing a coin. We can apply statistics such as calculating the average value which, for a great number of realizations, will give half heads and half tails, assuming the coin to be without flaws.

object of our study. One lies in the fact that we have to hypothesize about stochastic (random, and subject to chance) processes at work within the cosmos. What does this mean? Consider, for example, the range of luminosities of galaxies once they are formed. It is easy to imagine that everywhere, given similar physical conditions, the process of the formation of galaxies will lead to the appearance of similar, but not identical, objects. Their characteristics will be 'broadly' similar, rather like those of people of the same age range within the population of a given country. Of course, there will be differences from one country to another, as a result of their environments and living conditions. The same is true for galaxies, stars and planets. We must therefore seek statistical laws from which galaxies might be generated. In this game of 'pick a cosmic card', where we are trying to assess the probability of any particular card being drawn, we must also be aware of extreme examples (such as the brightest galaxies) and not see them as a *representative sample,*[2] just as basketball players are not representative of the height of people in general. Conversely, how do we interpret the presence in the real universe of, for example, a gigantic complex like the Virgo Cluster? Is it a 'normal' object, or the exceptional result of the evolution of our universe, which is, as we have stated, unique?[3]

From one to three dimensions

So, as we have said, there is only one solution: simulation. A simple enough aim: to recreate on a computer the evolution of the universe from its first moments until the present day, with a view to retracing the evolution of all the physical phenomena occurring within it. In practice, we will have to repeat the numerical experiment many times, trying to obtain the greatest possible number of realizations from initial conditions which vary but are compatible with what we know (since we do not know in detail what the real initial conditions were). From all these realizations we can deduce global properties which may be compared to our actual observations of the cosmos.

The first astrophysical simulations were carried out during the 1940s, when

---

2 A representative sample is a set of elements (things, people, data...) selected in such a way that it possesses the same properties as the population from which it is derived. For example, to try to forecast the result of elections, it is necessary to interview a range of people of different ages, sexes, places of residence and social strata.

3 Here, again, we mean the unique observable universe, which does not contradict the notion of the multiverse.

the first computers became available. In order to represent the interaction between two galaxies containing in reality billions of stars, just a few tens of points (stars) were employed in the interaction, and not in space but on a flat plane. We realize of course that the laws of gravitation allow us to calculate the positions of two bodies quite exactly (Kepler's famous Laws describe the orbits of the planets around the Sun), but as soon as three (or more) bodies are involved, there is no exact solution. Here is another good reason to turn to simulations! The limitation to two dimensions was due to the limited capacity of the computers of that period. The transition to three dimensions (i.e. real space) demands the introduction of a coordinate for the positions, velocities and accelerations of each particle, but most importantly obliges us to use a far greater number of particles to simulate a given region with the same quality of detail as in a calculation using fewer dimensions. The well known 'Moore's Law', which suggests the doubling of computing capacity every eighteen months (at least in recent years), allied to the use of several computers in parallel, has meant that we can now simulate the gravitational interactions of billions of particles in three dimensions, as is happening now in projects such as Millennium, Horizon, and Mare Nostrum.

Let us now go into more detail on how to proceed.

Tame dark matter

As we have seen, it is dark matter that dominates the totality of the mass of the universe. On the cosmic scale, it is gravity that leads the dance, being the only force able to affect this dark matter. So before we can represent the evolution of galaxies or the formation of stars, and so on, it is essential to describe the backcloth to the cosmic stage: dark matter structures. *A priori*, a simple task: the laws of gravitation are exactly known, thanks to Newton and, later, Einstein, who have explained them to us in detail, and there is no other interaction; we can generally ignore the effect of ordinary matter on dark matter.

However, gravitational interaction involves no screening effect, unlike that found in plasma in the case of electromagnetic interactions, even though the latter also theoretically work at infinite range. In fact, beyond a certain scale, ions and electrons of opposing charges organize themselves in such a way that the plasma may be considered as locally neutral, preventing the local electrical field of more remote particles acting upon the small volume in question. In the case of gravity, there is no equivalent to these opposing electrical charges, and each particle is therefore affected by all the others, whatever the distance between them may be. Moreover, the influence of a very large remote structure may be greater than that of isolated, more local particles. So when we calculate the new position of a given particle, we have to take into account *all* the forces created by billions of its fellows and apply the laws of dynamics (which tell us that acceleration is proportional to force, leading us to the deduction of velocities and then positions), for every single particle in the simulation.

The number of operations needed is therefore proportional to the square of the number of particles (the forces exerted by each particle on all the others). One initial problem is that dark matter is made of elementary particles whose exact mass remains unknown, although it cannot exceed 10^{-29} kg – and we have the task of reproducing the evolution of this matter on scales of millions of light years, involving masses of the order of tens of millions of billions of times greater than that of the Sun. So we have to simulate the evolution of more than 10^{80} particles, and all this on a time scale of 14 billion years. Even with the most powerful computers, the best of which are capable of a thousand billion elementary operations per second, we would need millions of years of calculation to simulate a significant part of the universe (or even one galaxy)!

So we have to fall back upon approximation, and bring the dark matter particles together as 'superparticles'. The mass of these superparticles may vary from a few thousandths of a solar mass, for simulations involving formation of the earliest stars, to approximately one billion solar masses, which is about the size of a dwarf galaxy, for simulations of the large-scale structure of the universe. So we are far from the notion of the elementary particle and one consequence is that we now have only a rough view of the universe, as if we were trying to represent the flow of a pile of sand from its total mass, but substituting boulders for the grains. The essential thing is not only that we know this, but also that we know which properties are correctly represented and which will not give us any pertinent information. For example, a 'superparticle' of one billion solar masses allows us correctly to represent galaxies weighing one-tenth of the mass of the Milky Way. However, dwarf galaxies such as the Magellanic Clouds, the two satellite galaxies of the Milky Way, will not be represented in such a simulation.

A second problem involves one of the fundamental parameters of simulation, the so-called 'time step', i.e. the interval between two successive realizations of our numerical universe (since computers cannot represent continuous phenomena). Our interval must not be too short, leading us to unrealistic calculational periods, since each interval implies the completion of calculations of gravitational forces already mentioned. However, if the interval is too long, the approximations of the calculations would be such as greatly to affect the reliability of the result. Given these constraints, time steps of about 2,000 years are used in large cosmological simulations. During this interval, a galaxy within a cluster will travel about ten light years, sufficient to modify the values of the gravitational forces to which it is subject. For simulations involving smaller-scale phenomena, such as star formation and the dynamics of galaxies, the values will be smaller. Now we approach the dangerous rocks, of which we have already been warned: we are now dealing with close interactions (even though only the simple law of gravitation is yet involved). It is well known that the force of gravity between two particles increases rapidly as the distance between them decreases, the force F varying as $1/r^2$, i.e. as the inverse of the square of the distance between them. So when we deal with particles that are close to each other, the small distances involved imply greater force, and greater acceleration (recalling the famous law of physics we learned at school $F = ma$, linking force to acceleration), and this varies rapidly.

The difficulty will lie in following these large variations precisely through time, across shorter and shorter intervals, while a computer, by definition, can only calculate at separate and finite moments. It is therefore indispensable in practice to use an interval of time adapted to acceleration, and therefore to lessen it in the case of interactions at small distances. This is inconvenient, because it slows our simulation, and we must not allow our calculation time to approach the age of the universe. Another possibility is to have recourse to (controlled!) approximations of these situations: in effect, if two particles are very close to each other, it is possible to ignore the influence of all the others. So the trajectories of two interacting bodies are exactly calculable. We can therefore include these calculations in the simulation if we detect two particles that are too close to each other.

Bearing all these precautions in mind, we set to work. The first large cosmological simulation was that known as the Millennium Run, carried out by a German team at the Max Planck Institute in Garching, with input from British, Canadian, Japanese and American collaborators. This simulation involved ten billion particles in three dimensions and employed, for a month, a computer with 512 processors, corresponding to a total of 42 years of calculation. (The processor is the 'intelligent' part of the computer system that carries out the instructions of the program. The more processors a computer has, the more operations it can deal with simultaneously.)

During the simulation, part of the computer program (algorithm) identifies structures in the process of being created, recording 'gatherings' of superparticles. The smallest galaxies are identified as groupings of twenty superparticles, corresponding to a galaxy ten times smaller than the Milky Way. A cluster of galaxies may contain more than 10,000 particles. The initial conditions of the simulation, i.e. the initial values of the various physical quantities (such as density and temperature) of the cosmic fluid are chosen in order to take into account the observation of temperature fluctuations within the cosmic microwave background, which determines the distribution of the density fluctuations in the dark matter which we wish to track over time.

It then remains to select the initial moment from which the simulation will begin, and press 'return' on the keyboard.

The need for semi-analytical methods

Numerical simulation therefore allows us to make the initial density fluctuations evolve according to the various physical hypotheses (scenarios) which we wish to test. Our ultimate constraint is of course that our results should correctly mirror the body of observational data obtained by telescopes, at various redshifts and wavelength domains.

It is a complicated undertaking, certainly not as easy as might at first be imagined. Initially, it seems simple enough to follow the evolution of dark matter and baryons subjected to gravity (i.e. before the birth of the earliest stars), but subsequently, things become much more complicated. As soon as a galaxy forms, its density is much greater than that of intergalactic gas, and it will modify the gravitational field around it. Even if baryons represent only a fifth of the total mass, their action cannot always be ignored. What is more, the stars emit light, which can modify the state of ionization (i.e. more or less stripping electrons from atoms) of the surrounding gas, and that at great distances. Also, quasars, fewer in number but very luminous, have a major effect upon the intergalactic medium. Finally, exploding stars end their lives, releasing heavy elements, synthesized in their hearts, at great velocities into the interstellar medium. Now, these phenomena may occur on scales far smaller than that of our superparticles (for example, star formation), or may involve multiple sources whose individual contributions are impossible to assess. They must be calculated broadly, e.g. the flux from quasars, an indispensable ingredient in the calculation of the degree of ionization of the intergalactic medium. In all cases, it is impossible exactly to simulate all these complex phenomena, for which we may not even have all the details.

To overcome these difficulties and the impossibility of taking into account all phenomena and their details, we resort to so-called 'SAMs' (semi-analytical methods) based on observations of these phenomena. For example, it appears that the mass distribution of stars that have formed in a galaxy is always roughly the same, in the same manner as the rate of formation, and we can 'approximate' them with laws. So when, in a simulation the chosen algorithm identifies that a structure the size of a galaxy is decoupling from the expansion and is of sufficient density for star formation to proceed, it deduces, using pre-programmed formulas, the proportion of baryons transformed into stars, and the intensity of the emitted radiation. The time saved in calculating is indeed considerable and the validity of the approximation borne out by comparison with specific observations. If necessary, we can change the model for the phenomenon in question, proceeding pragmatically.

Fiat lux[4]

However, the numerical cosmologists still have work to do: especially if their aim is to identify baryons, which, everything seems to indicate, are hidden in the intergalactic medium. This gas is in fact not really primordial, but reveals traces of contamination by heavy elements (mostly carbon and oxygen). How can these elements, which can have been formed only in the cores of stars, have found their way into intergalactic clouds hundreds of millions of light years away from the nearest galaxies? The only viable hypothesis is that they have come from supernovae, colossal explosions signaling the end of the lives of the most massive stars: according to simulations of the gas surrounding them within their host galaxies, they can eject material at very high velocities. This can send gas enriched with heavy elements very far away, escaping from the gravitational field of the galaxy into the intergalactic medium. Here too, simulation is not easy: obviously, we cannot reproduce in whole-universe simulations the action of each supernova. On the other hand, detailed simulations of these objects (alone requiring as many calculations as our whole-universe simulations, since the physical phenomena occurring in supernova explosions are complex), allow us to work out with reasonable accuracy the amount of energy liberated, the quantities and composition of the ejected material, or the luminous flux. Further simulations, on the scale of a galaxy, will lead to the deduction of the quantity of matter that such explosions hurl out of galaxies, polluting the primordial intergalactic medium. These values are then integrated into our simulation of the universe, into which we have also programmed the rate of supernova explosions per galaxy.

In this way we can reproduce a complete map (of density, chemical composition and temperature) of the intergalactic medium, by combining information on the density of the gas (through gravitation), the luminous flux of quasars and supernova explosions (Figure 7.1). Mission accomplished! All that now needs to be done is to pass the computer-simulated portion of the universe through an imaginary telescopic line of sight, identify each cloud traversed, and extract its physical parameters in order to evaluate the absorptions detectable in the spectrum of a quasar, at whatever redshift this might be. From these virtual measurements, we can then reproduce observations of real quasars, verifying that the simulation correctly reproduces the characteristics of the Lyman-alpha forest.

The most recent calculations reproduce these observations, and the global properties of galaxy distribution, very accurately, in terms of both space and luminosity. A few difficulties remain, especially in the case of the faintest

---

[4] "let there be light"

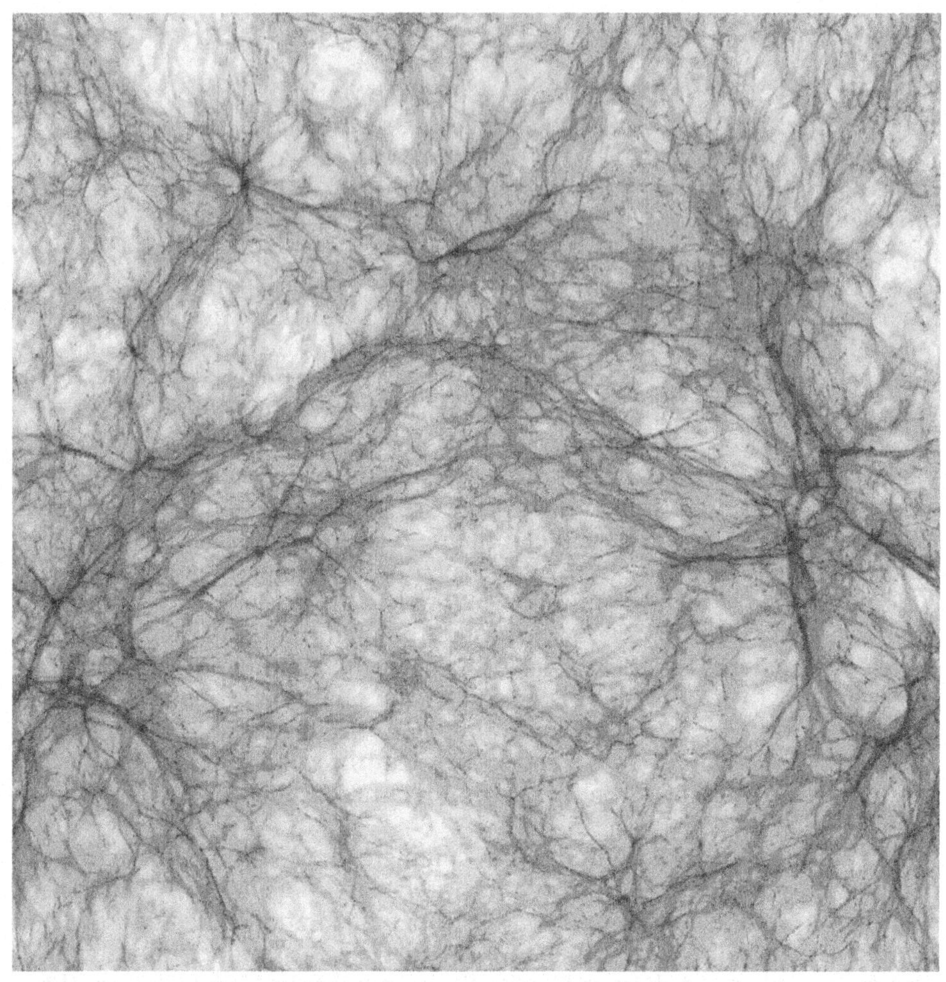

Figure 7.1 Simulation of the distribution of baryons. The map represents a cube-shaped portion of the universe 150 million light years on a side. The filaments correspond to the distribution of gas at z = 2.5. This calculation required a computer capacity of 2,048 processors (Project Horizon).

galaxies (which is hardly surprising, since we are working at the limits of the simulation's 'resolution'), and also of high-redshift galaxies. The latter, which have merged to form the most massive modern galaxies, seem in fact more numerous in observations than in simulations. The broad fact remains that it is now possible correctly to reproduce the history of the universe on a computer.

Simulating the universe in computers

The universe and the objects in it are three-dimensional, so simulations of them require us *a priori* to take all the dimensions into account. There are, however, problems which do not necessarily require complete simulations. Simulating the interior of a star, for example, can at first approximation be carried out in only one dimension, assuming spherical symmetry. However, it soon becomes apparent that the rotation of the star will introduce different behaviors according to the direction in question. The presence of convective cells (enormous fluxes of matter circulating between the interior and the surface, transporting heat exactly like water boiling in a heated saucepan when the quantity of heat to be evacuated becomes too great) requires, in the case of detailed study, that the simulation move into two or even three dimensions.

The same is true for the development of structures of matter in the universe. If we at first suppose that a given structure is spherical, working in one dimension will suffice to study its evolution, for example identifying the period at which this structure will decouple from the expansion of the universe. (In fact, at the beginning of their growth, structures can continue to increase in density while they are subject to expansion: all that is necessary is for the accumulation of matter to compensate for the dilution.) But when the quantity of matter becomes too great, the structure begins to collapse in upon itself and gravity overcomes expansion: a 'self-gravitating' structure is the outcome.

This is the case with galaxies or clusters of galaxies, systems bound by their own gravity, their internal structure no longer subject to the expansion of the universe. Conversely, filaments and great voids, the least dense areas on average, will grow in tune with the expansion. Knowing at what epoch the dark halos, progenitors of galaxies, condensed is crucial to our understanding of the history of the formation of stars, and therefore of baryons, in the universe.

As distance decreases, the force of gravity increases and acceleration increases. Now, acceleration is the temporal derivative of the velocity of the particles, which is itself a temporal derivative of position. However, in simulations, only successive positions are available, separated by the same time intervals. Let us suppose that a particle has a constant acceleration of 100 meters per second^2 (i.e. an additional 100 meters per second every second), its initial velocity is nil, and the time interval of the simulation is one second. Now, the velocity after one second is 100 meters per second, and the particle will not have moved (its initial velocity being nil). After two seconds, it will have travelled 100 m, and its velocity will be 200 meters per second. After three seconds, it will have travelled 300 (= 100 + 200) meters, and its velocity will be 300 meters per second. And after four seconds, its velocity will be 400 m/s, and it will have travelled 600 meters. If we now use a two-second time step, after the first step the particle will not have moved, and its velocity will be 200 meters per second. After the second step, i.e. after 4 seconds, it will have travelled 400 meters and its velocity will be 400 meters per second. So we see that, although the velocity is correct in this simple example where the forces are constant, the position is now 33 percent in error. To take account of this rapid evolution of acceleration, the time step would have to be constantly revised. What we gain in temporal 'resolution' and in the accuracy of positions and velocities serves to drastically slow down and lengthen the calculation.

Any evolutionary calculation requires a starting point, its 'initial conditions'. Think of the classic experiment in the science class, involving a marble and an inclined plane: the initial position and velocity of the marble, and the inclination of the plane, must be known in order to determine the trajectory. In the case of the universe, as well as the values for physical parameters (density, etc.), we have to nominate a starting point (i.e. redshift z). This will probably be the time of the recombination (z = 1,000), since the physical parameters then in place are those that the cosmologist knows best. But in practice, until the earliest stars shone, gravity held sway as the density fluctuations which would eventually form galaxies grew. To the great advantage of cosmologists, all the while structures remain tenuous, they can perform analytical calculations, i.e. find exact solutions for the evolution of contrasts in density. Once these have become too great, this is no longer possible, and the aim of the simulations is to study this phase. So we can save cosmic time as we calculate by choosing as our starting point a redshift z of about 100, as was done with the Millennium Run project.

Another problem, this one technical, involved the management of these numerical simulations. Apart from the capacity required of the computers (1 TeraOctet of active memory: 1 To = 1,000 Go, or the equivalent of the active memory of 500 PCs), there is also the problem of data storage. For each time step and for each particle there is a finite quantity of information (position, velocity, etc.), but we are dealing with billions of these particles and if we introduce, for example, 10,000 time intervals, numbers become 'astronomical'. So the simulator is faced with a dilemma: storing all the data acquired is not easy. What should be kept? Which of these calculations, which have cost the scientist so much in time and money, should be disposed of? In the case of the Millennium Run project (Figure 7.2), only 64 runs were retained, requiring 20 TeraOctets of disk space (the equivalent of about 100 PC hard disks).

When we have finished our calculations, there is one more procedure to be carried out before we can compare the simulation with observations: 'simulating the observations', since, no matter how exacting the observations have been, they distort reality because of flaws in the instruments or detectors, and because of their physical limitations (in resolution, wavelength, sensitivity, etc.). So these flaws must be reproduced and applied to the data provided by the simulation. To do this, we require data to calibrate the instruments precisely, or a simulation (yet another!) of these

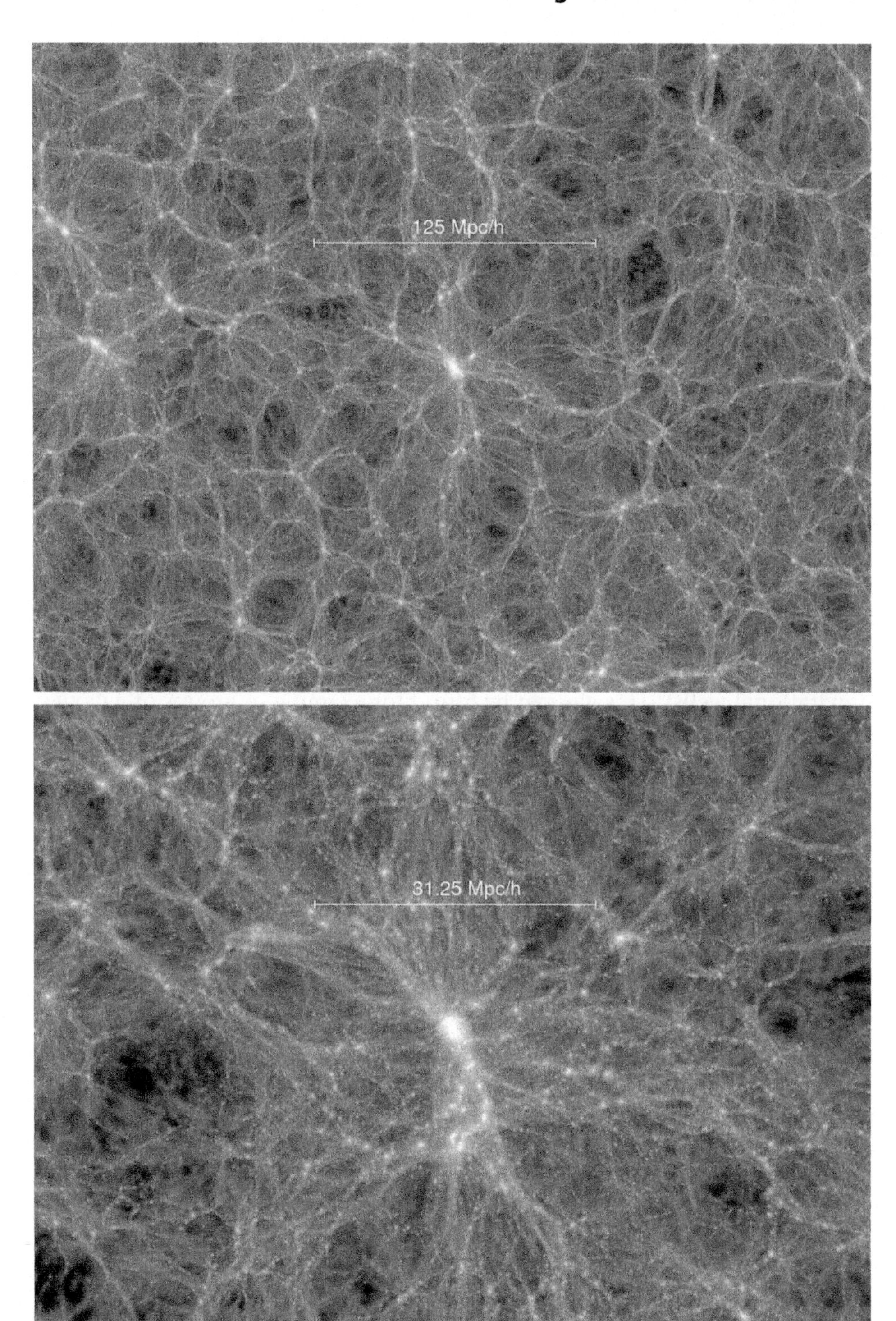
125 Mpc/h
31.25 Mpc/h

instruments. All that then remains is to analyze the data from our virtual instruments and compare them to the results obtained by real observations of the cosmos in order to verify the validity of the simulation, an indispensable condition to be able to use the results to predict new effects or prepare new observations. We should however note that this condition is not totally sufficient: it is not because the observations of that which we are capable of observing are coherent with the simulation that it reproduces correctly everything in the universe. At most it is an indication that the simulation is globally correct. But any disagreement will only serve to further our knowledge of the cosmos.

Baryons: pinned down at last

If we trust the ability of simulations to reproduce the evolution of baryons, we can use these calculations to find out where the still missing ~50 percent are hidden.

What the simulations show is that part of the gas is to be found in an intermediate state between the Lyman-alpha forest (corresponding to gas at a temperature of 10,000 degrees, heated by the intense ultraviolet radiation from quasars) and the gas of clusters of galaxies (heated to more than one million degrees by collisions between atoms trapped in the clusters' gravitational wells). This gas can be found for example in groups of galaxies, which are less massive than clusters, or in the exterior regions of clusters, which are cooler than the X-ray emitting centers.

Figure 7.2 The Millennium Run simulation used more than 10 billion particles to trace the evolution of the matter distribution in a cubic region of the Universe over 2 billion light years on a side. It kept busy the principal supercomputer at the Max Planck Institut's Supercomputing Center in Garching, Germany for more than a month. By applying sophisticated modeling techniques to the 25 terabytes of stored output, Virgo Consortium scientists have been able to recreate evolutionary histories both for the 20 million or so galaxies which populate this enormous volume and for the supermassive black holes which occasionally power quasars at their hearts. By comparing such simulated data to large observational surveys, one can clarify the physical processes underlying the buildup of real galaxies and black holes. Here we see two frames from the simulation showing a projected density field for a 15 Mpc/h thick slice of the redshift z = 0 (t = 13.6 Gyr) output (top) and the z = 1.4 (t = 4.7 Gyr) output (bottom).

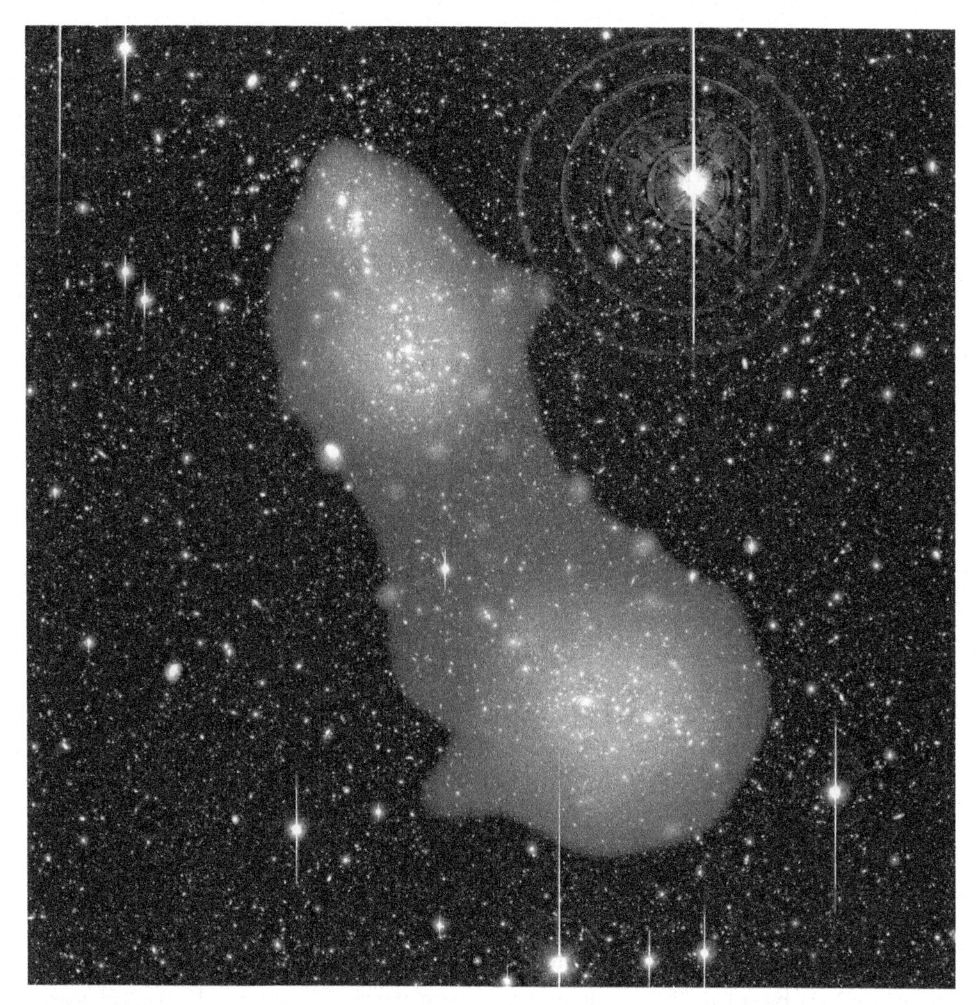

Figure 7.3 Superposition of the optical image (taken with the Suprime-Cam of the Subaru telescope) and the X-ray image (ESA XMM-Newton) of the pair of clusters Abell 222 and Abell 223, 2,300 million light years away. The two yellow areas correspond to X-ray emissions from hot intracluster gas at several million degrees, characteristic of these massive systems. A filament connecting these two main areas of emission was discovered, corresponding to a gas at a lower temperature (darker red). This filament represents the 'missing baryons' that are the subject of this book (ESA/XMM-Newton/EPIC/ESO (J. Dietrich)/SRON (N. Werner)/MPE (A. Finoguenov)).

But why have these baryons not been detected, as have the others? Because the gas is at a temperature of around 100,000 degrees and so almost all of the hydrogen has been ionized. It therefore produces no absorption in the Lyman-alpha line, which only occurs in the case of (neutral) atoms. Moreover, it is not hot enough produce any detectable X-ray emissions. Finally, even if this gas contains traces of elements other than hydrogen and helium, its high temperature implies that they will be highly ionized, and therefore stripped of electrons. All that is left in those elements are energy levels for which the photons emitted are in that part of the X-ray domain that is difficult to observe. The same is true for all lines that the elements in the gas could produce. This 'soft' X-ray domain is not in fact accessible to optical or ultraviolet instruments, the technology not yet being adapted for it. This is why this gas has remained 'unseen' by observers. Since the other contributions to the baryon total have only recently been identified with any certainty, the amount of this gas was very uncertain, reducing the likelihood that instruments dedicated to its discovery would emerge. However, interest in this unidentified baryonic matter has once again become lively (see next chapter), in the light of the results that we have described throughout this book so far.

Now, observatories like NASA's Chandra and ESA's XMM-Newton have soft-X-ray spectroscopic capability, working at wavelengths of the highly ionized gas revealed by simulations. They are also sensitive enough to detect gas that is cooler and less dense than that found at the heart of clusters. For example, XMM has been able to detect a gigantic filament connecting two clusters of galaxies (Figure 7.3) with characteristics predicted by simulations. The power of the numerical method as a predictive tool is thereby reconfirmed. As far as absorptions by more ionized elements (for example oxygen or neon) are concerned, results are still uncertain, for these lie at the limit of sensitivity of the most modern instruments. Nevertheless, all the results show that this gas contains in fact all the missing baryons, and our enquiry can be closed. So now the accounts are balanced. It is finally possible to reconstitute the history of ordinary matter, which is all around us and is our very substance, through all cosmic time, from the birth of the elements to the present day (Figure 7.4). Another resounding success for the cosmological model!

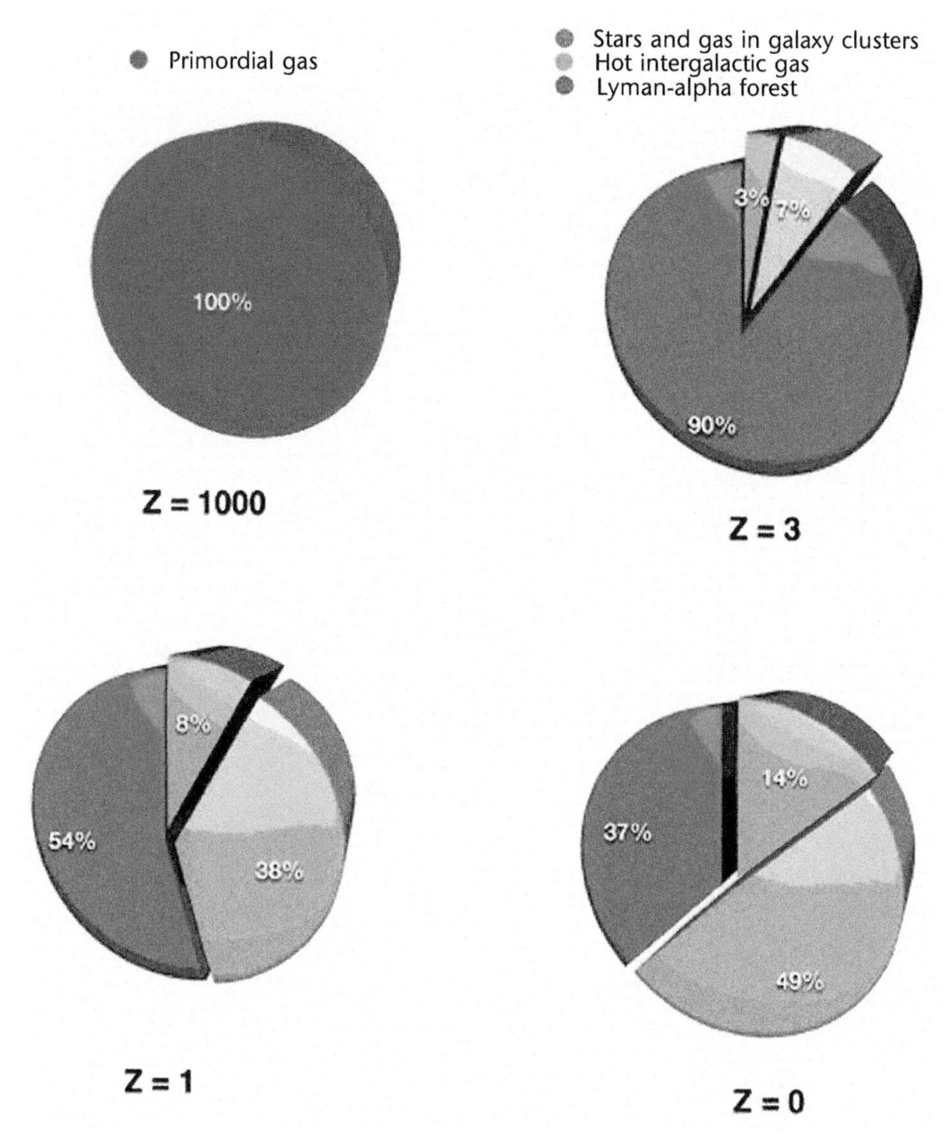

Figure 7.4 The evolution of baryons through time. The evolution from z = 1,000 to z = 0 of the various phases of the history of ordinary matter. Primordial nucleosynthesis had fixed, by the end of the first three minutes of the universe, the total quantity of ordinary matter. Observation of the harmonics of the cosmic microwave background reveals the totality of baryons, still in the form of near-homogeneous hydrogen-helium gas, at z = 1,000. While all is accounted for at z = 3, this is no longer the case at z = 0 (present-day universe). By means of simulations and dedicated instruments, we have been able to 'lift the lid' on the hidden baryons. We can also show them for any given epoch, such as z = 1, corresponding to approximately half the age of the universe.

8 A constant search

"A search always starts with beginner's luck and ends with the test of the conqueror." Paulo Coelho

Progress in science is accompanied (and more often than not, preceded) by major technological advances. Throughout this book we have seen how the two are linked: the previous chapter showed us the impact that dazzling improvements in information technology have brought about. The last chapters will review other advances and some promising projects now under way.

The optical sky: from naked eye to CCD

The Greek astronomer Hipparchus may be considered the father of quantitative astronomy: not only did he undertake the major task of cataloguing around 850 naked-eye stars, but he also introduced the notion of magnitude, classifying stars in terms of their apparent brightness.[1] Not until the seventeenth century did the first astronomical instruments (telescopes) supersede the human eye as detectors, increasing at a stroke the domain of the observable universe.

An inherent difficulty with astronomical instruments is that normally they allow us to see only a restricted area of sky: their fields of view (the area of sky effectively observed through the instrument) are small. For example, the field of view of the Hubble Space Telescope, an instrument capable of observing the most remote galaxies yet detected, is equal to one-ninth the apparent diameter of the full Moon, equivalent to the size of a pinhead held at arm's length (Figure 8.1).

The Milky Way, straddling the sky in summer, was for a long time the only region accessible to astronomers studying individual stars, and is far too large to observe all at once with most telescopes. The invention of the Schmidt camera, the first of which was built in 1930, offered a partial solution, with a

---

1 Hipparchus assigned the brightest stars a magnitude of 1, the next brightest magnitude 2 and so on to the faintest stars, just visible to the naked eye, which were magnitude 6.

Figure 8.1 Fields of view. The Hubble Space Telescope's Advanced Camera for Surveys (ACS) and Wide Field and Planetary Camera (WFPC) have fields of view of, 3.4 and 2.7 minutes of arc, respectively, about one-ninth of the diameter of the full Moon (NASA/STScI).

field of view of several square degrees (the whole sky covers 40,000 square degrees). This type of wide-field instrument is suited to programs of observation where large numbers of objects (stars and galaxies) or moving objects (comets, asteroids, artificial satellites) are to be studied. The best known of these instruments is the 48-inch (1.22-meter) aperture Schmidt camera (the Samuel Oschin telescope), at the Mount Palomar Observatory in the USA (Figure 8.2), able to image one thousandth of the sky at one time. It was used for the first Palomar Observatory Sky Survey (POSS), which mapped

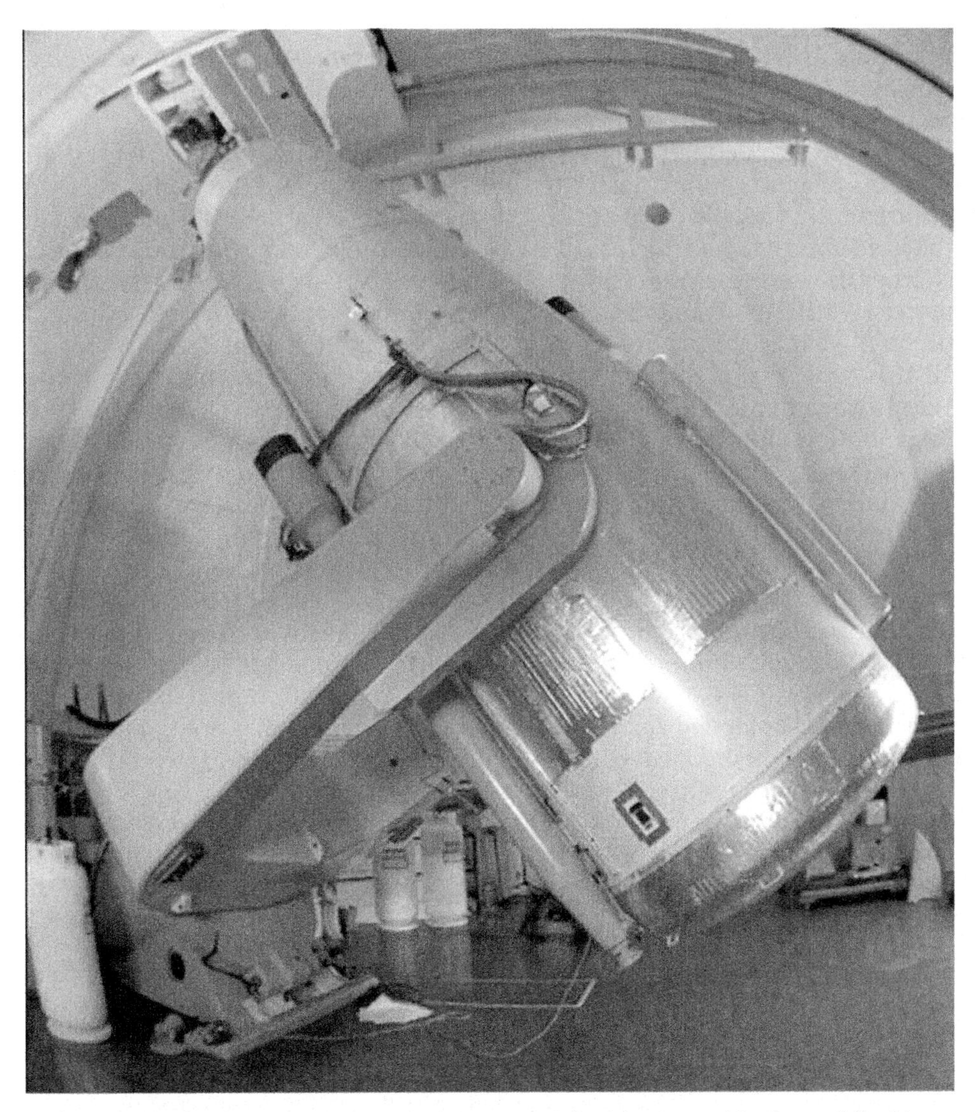

Figure 8.2 The 48-inch (1.22-meter) aperture Schmidt camera (the Samuel Oschin telescope), at the Mount Palomar Observatory in the USA. It was used for the first Palomar Observatory Sky Survey (POSS), which mapped the northern hemisphere of the sky during the 1950s (Palomar Observatory).

the northern hemisphere of the sky during the 1950s. This catalog was for a long time the basic workbook for researchers in extragalactic astrophysics. A similar instrument in southern France, on the Plateau de Calern, has been searching for asteroids for some time now. These telescopes are nowadays equipped with spectrographs capable of observing hundreds of stars simultaneously, such as the Two Degree Field system ('2dF') at the Australian Astronomical Observatory,[2] or with very large CCD arrays which have done away with the need for photographic plates.

Most modern people snapping away with their camera-phones, with their millions of pixels,[3] may have little idea of what the era of emulsion-based photography was like. For about 150 years, this was the only method, all that astronomers had to record and catalog their images of the universe. Neither they nor the professional photographers had an easy time of it: ever more sensitive emulsions had to be invented to record fainter objects and to probe the various domains of the visible spectrum (from violet to red), and all this on a wide field that was even (slightly) curved at the focus of the Schmidt camera. Moreover, once the exposure was completed, the astronomer would have to carefully retrieve the glass plate, about 30 cm square, for developing. The quantity of information on these exceptionally high-quality photographic plates was certainly considerable, but extracting that information and drawing scientific conclusions from it involved yet more painstaking work.

The quantity of energy received from a celestial source, translated into the darkening of the emulsion of the photographic plate, had to be estimated to extract the required information, so the image obtained had to be digitized to find the brightness of the object observed. Digitization is the process of transforming analogue data into digital data. In digital photography, the image is obtained/coded directly onto the pixels, each if which is an individual receiver. In the case of old-style photography, a machine was required to digitize an image, i.e. represent it as a series of numbers on a grid of pixels. Each pixel value was characteristic of the intensity of the signal received at that spot. However, the relationship between the quantities

---

[2] The Two Degree Field system ('2dF') is one of the world's most complex astronomical instruments. It is designed to allow the acquisition of up to 392 simultaneous spectra of objects anywhere within a two degree field on the sky. It consists of a wide field corrector, an atmospheric dispersion compensator, a robot gantry which positions optical fibers to 0.3" on the sky. The fibers are fed to a duel-beam bench mounted spectrograph stationed in the thermally stable environment of one of the telescope Coudé rooms. A tumbling mechanism with two field plates allows the next field to be configured while the current field is being observed.

[3] A digital image is composed of pixels, a contraction of *pix* ('pictures') and *el* (for 'element'). Each pixel contains a series of numbers which describe its color or intensity. The precision to which a pixel can specify color is called its bit or color depth. The more pixels the image contains, generally the better the resolution, and the more detail it has the ability to describe. A 'megapixel' is simply a million pixels. Modern digital cameras contain sensors with 16-21 megapixels.

involved was not a direct one, and an additional phase, that of calibration, was required. This was a particularly delicate and complex operation, requiring the setting up of facilities with specialized machines such as the UK's Automated Plate Machine (APM) and France's Machine à Mesurer Automatique (MAMA), both now defunct, for the digitization of the sky survey results. At the time, information technology was far less advanced than it is nowadays (remember Moore's Law, which states that computing power doubles every 18 months). In spite of all these difficulties, useful results were obtained such as the mapping of millions of galaxies by the APM group (Figure 8.3).

Figure 8.3 The Automated Plate Measurement (APM) Galaxy Survey was one of two surveys (the other being the Edinburgh-Durham Southern Galaxy Catalogue) carried out in the late 1980s and early 1990s which digitized hundreds of sky survey plates in order to compile a catalog of galaxies over a very wide area of sky. The survey plates were taken with the UK Schmidt Telescope in Australia. Each plate covers an area of six by six degrees on the sky and the plates are spaced on a uniform grid at five degrees apart. The one degree of overlap between neighboring plates proved vital for calibrating the individual plates onto a uniform magnitude system. The plates were individually scanned using the Automated Plate Measuring (APM) facility, a high-speed laser microdensitometer, in Cambridge, England. Astronomical sources are detected in real time as the plate is scanned and the position, integrated density, size and shape of each source is recorded. Galaxies and stars are distinguished by their shapes: galaxies are fuzzy, extended sources whereas stars are much more concentrated (APM Galaxy Survey).

Although the use of large-format photographic plates on Schmidt telescopes solved the problem of restricted fields of view, the subsequent, analytical steps brought with them several difficulties; for example the non-linear response of the plate to the radiation received, and saturation of the emulsion by very bright stars, which remained definite obstacles. The appearance of charge-coupled devices (CCDs) in the late 1960s caused a revolution in astronomical observation. These instruments, composed of grids of pixels did away with subsequent digitization. Photons striking the detector strip electrons from the (silicon-based) material, and these are captured by a positive electrode. All that is then needed is to read the content of each element by successively emptying them (as the electrons leave they create a current proportional to their numbers). The intensity of this current is then translated into a number stored on a computer. The total reading time is of the order of one minute, after which the observer has a file containing the image. Teething troubles such as their limited format (e.g. 50,000 pixels) and limitations in accessible wavelengths have all now been completely overcome.

CCDs are now found not only in mobile phones, PDAs, webcams, optical scanners and other digital devices, but also in impressive astronomical cameras such as the MegaCam mounted on the 3.6-meter Canada-France-Hawaii Telescope in Hawaii (Figure 8.4), which has a mosaic of 36 CCDs (each of 2,048 4,612 pixels) working in the visible and near-infrared. A similar instrument is also operating on the 8.2-meter Subaru Telescope, also in Hawaii. Projects are under way to construct even larger cameras for both ground-based and space platforms. An example is that of the Large Synoptic Survey Telescope (LSST), with a three-billion-pixel camera consisting of a mosaic of 16-million-pixel detectors at the focus of a ground-based 8.4-meter telescope (Figure 8.5). As for space-based instruments, detectors of about 500 million pixels are foreseen within the next decade, in particular for the measurement of cosmological constants. The (very) big detectors are certainly on the march.

Figure 8.4 (Top) The 3.6-meter Canada-France-Hawaii Telescope (CFHT) in Hawaii, shown here at night with star trails behind (Jean-Charles Cuillandre (CFHT)). (Bottom) Mounted on this telescope is the wide-field camera, known as MegaCam, which covers about 1 square degree of sky, and is at present one of the largest imagers in the world. It has a mosaic of 36 CCDs (each of 2,048 $\times$ 4,612 pixels) working in the visible and near-infrared (Paris Supernova Cosmology Group).

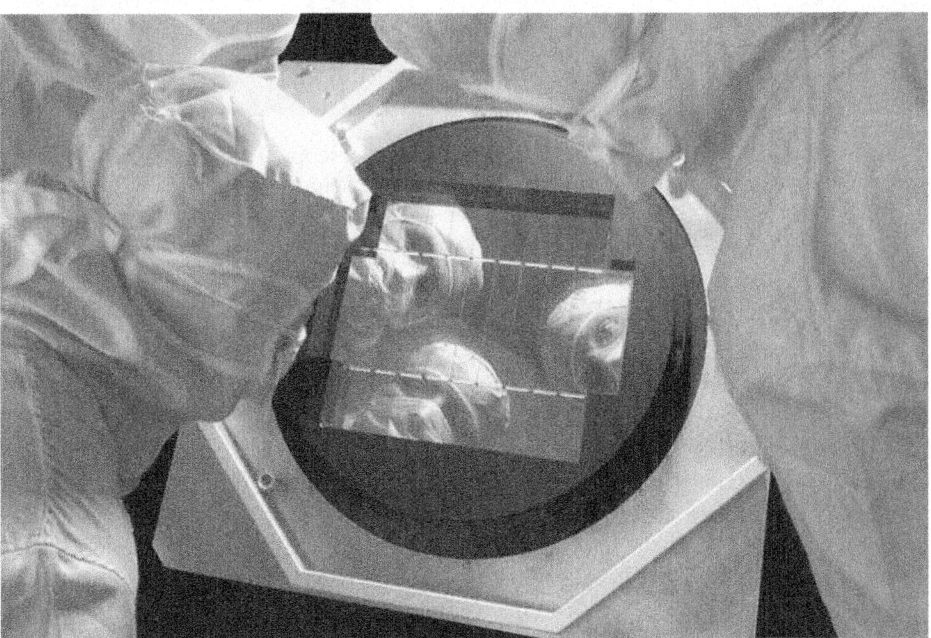

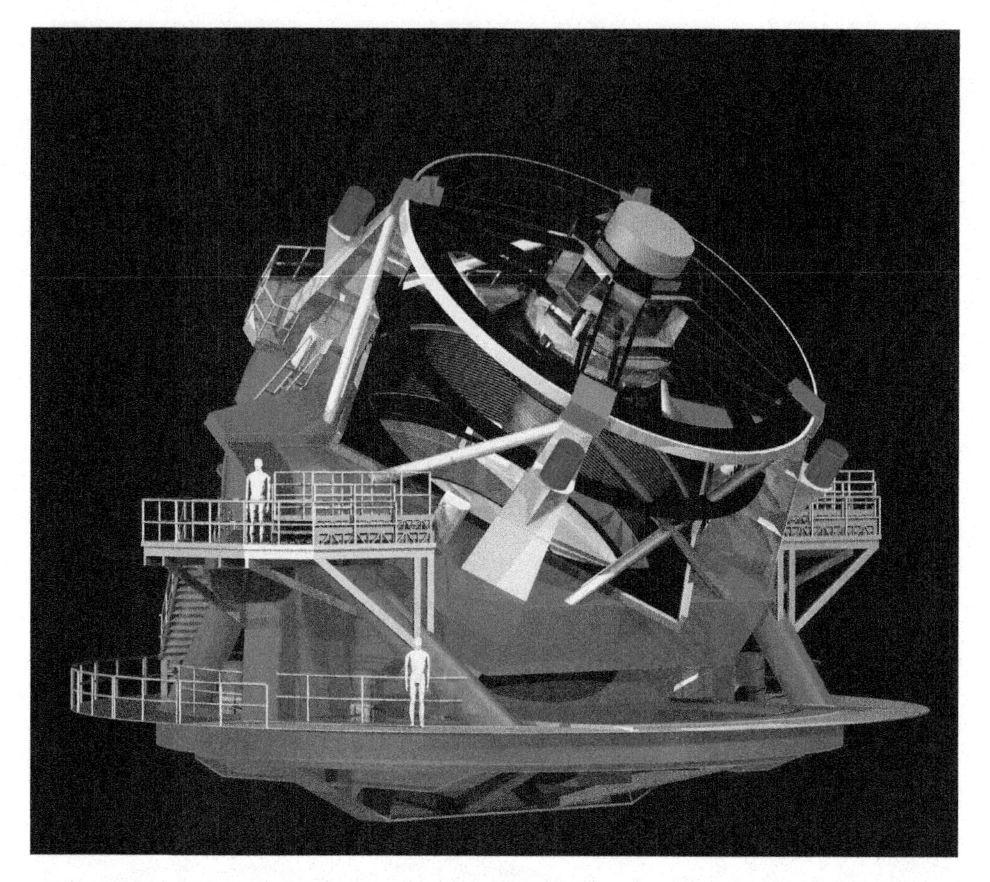

Figure 8.5 The 8.4-meter Large Synoptic Survey Telescope (LSST) will use a special three-mirror design, creating an exceptionally wide field of view and will have the ability to survey the entire sky in only three nights (LSST Corporation).

Space odyssey

In chapter 3, we discussed the close links between advances in spaceflight and the ensuing momentous and often unexpected progress in astronomy. One striking example is the discovery by a spy satellite in 1967 of gamma-ray bursts, some of the brightest objects in the universe, but radiating only in the gamma-ray domain. Other, more planned advances have been and will be possible as we make the best use of our capabilities for space-based observation. No longer do we have to put up with the distorting effects of the atmosphere, the vagaries of the weather, and the glare of the full Moon.

The European Space Agency (ESA) rode this wave with the Hipparcos astrometry mission dedicated to the precise measurement of star positions (described in detail in chapter 1). Hipparcos' measurements played an essential role in many fields of astrophysics, including cosmology. One of its benefits was that it led to a better understanding of the shape of the Milky Way, and the evolutionary contexts of stars. New distance measurements gave more information about Cepheids, variable stars whose periods are related directly to their luminosities, meaning that they can therefore be used as 'standard candles' or cosmic yardsticks.[4] The only complication is that Cepheid variables are divided into several subclasses which exhibit markedly different masses, ages, and evolutionary histories. These have different relationships between their luminosity and period. Failure to distinguish these different subclasses in the early-to-mid 20th century resulted in significant problems with the astronomical distance scale. Therefore, it is necessary to find out which type of Cepheid variable is being observed before this method can be used.

This technique led to the modification of our estimates for the age of the universe and of the Hubble Constant. Following on from the success of Hipparcos, ESA is now working on its successor, the Gaia project. The aim of Gaia, due for launch in 2013 (Figure 8.6), is to establish a three-dimensional map of the Galaxy, not only through precise positional measurements (to microseconds of arc) but also of velocities, measured by spectroscopy, for about one billion stars. These exceptional data will deepen our knowledge of the composition, formation and evolution of the Milky Way. (The Gaia project is described in more detail in chapter 1.)

The X-ray domain, discovered by German physicist Wilhelm Roentgen in the late nineteenth century, was for a long time inaccessible to astrophysicists, because our atmosphere screens out all such radiations. To overcome this problem, we must observe from space, a strategy that became possible only after the Second World War. The earliest such scientific observations were carried out either by rockets, with flight times of only a few minutes at most, or by balloons, which stayed up longer, but could not reach the same altitudes as the rockets, thereby rendering them less efficient in detecting radiation in the X-ray domain. Only satellites could provide both duration and capability of observation.

The Sun is our local star, the major source of energy on Earth, without which life on Earth would not exist. It emits not only the visible radiation (sunlight) to which we wake each morning, but its infrared radiation produces the feeling of warmth on your skin on a sunny day, and its ultraviolet rays are responsible for the suntans of beach and mountain

---

[4] From the light curve of the Cepheid variable, its average apparent magnitude and its period in days may be determined. Knowing the period of the Cepheid we can determine its mean absolute magnitude from the appropriate period-luminosity relationship. Once both apparent magnitude and absolute magnitude are known, it is possible to work out the distance to the Cepheid.

Figure 8.6 An artist's impression of the Gaia spacecraft against the background stars of the Milky Way. Gaia will conduct a census of a thousand million stars in our Galaxy, monitoring each of its target stars about 70 times over a five-year period, precisely charting their positions, distances, movements, and changes in brightness (ESA, illustration by Medialab).

holidays. To the astronomer, the Sun is an 'ordinary' star, and is characterized as a dwarf. In 1949, an experiment aboard a recycled V2 rocket made the remarkable discovery that the Sun also emits X-rays. The OSO (Orbiting Solar Observatory) series of nine science satellites was intended to study the Sun at both ultraviolet and X-ray wavelengths; eight were launched successfully by NASA between 1962 and 1975. More recent ultraviolet and X-ray studies of the Sun have been carried out by SoHO (Solar and Heliospheric Observatory), a joint project of ESA and NASA, and by the Japanese spacecraft Yohkoh and Hinode.

After a long period of uncertainty in detection, a new milestone was reached when the existence of cosmic X-ray sources outside the Solar System was confirmed. The first such discovery, in 1962, was Scorpius X-1 (Sco X-1), whose X-ray emission is 10,000 times greater than its visual emission (whereas that of the Sun is about a million times less). Furthermore, the energy output of Sco X-1 in X-rays is 100,000 times greater than the total emission of the Sun across all wavelengths. Uhuru was the first satellite launched specifically for the purpose of X-ray astronomy; its mission lasted from 1970 to 1973. The Copernicus spacecraft (originally OAO-3) operated from 1972 until 1981 and made extensive X-ray observations. Among its significant discoveries were several long-period pulsars,[5] with rotation times of many minutes instead of the more typical one second or less. Home of possibly the most famous pulsar is the Crab Nebula (Figure 8.7), generally the strongest persistent X-ray source in the sky. This object is the remnant of a supernova seen to explode in 1054, the explosion having been so enormous that the object was visible in daylight for about three weeks.

The first extragalactic X-ray source to be identified was the galaxy M87 (Virgo A, 3C 274) in the Virgo cluster, in 1966. At about the same time, a diffuse high-energy X-ray background was detected, not identified with any particular star or galaxy, and the subject of a long-drawn-out debate. Later, other galaxies (particularly quasars) and clusters of galaxies were found to be emitting X-rays. The radiation from galaxy clusters comes not from the galaxies themselves but from very hot gas (at up to 100 million degrees). This gas is essentially left over from the gas (and dark matter) that formed the galaxies. The totality of dark matter, galaxies and gas constitutes a gravitational well of some 100,000 billion solar masses. Subject to this gravitation, of which they are also the cause, the galaxies and residual gas acquire energy and will finally reach a state of equilibrium. The presence of this considerable mass causes the galaxies to attain velocities of the order of several hundred kilometers per second, and the gas to reach temperatures of many millions of degrees (Figure 8.8). Various American, European and Japanese satellites have been launched to study these systems (e.g. EXOSAT (1983), ROSAT (1990), ASCA (formerly Astro-D, 1993), BeppoSax (1996), Chandra (1999), XMM-Newton (1999), and Suzaku (formerly Astro-E2, 2005)). The highly successful Chandra X-ray Observatory is shown in Figure 8.9.

One important result of all these experiments involves the quantity of baryons in clusters of galaxies, and the value of the characteristic cosmological parameter of the totality of matter (dark + ordinary) in the universe $\Omega_m \sim 30$ percent. One way of measuring this particular parameter is by studying a collision between clusters, allowing the possibility of the separate detection of baryonic and dark matter. At present, the future of the exploration of the cosmos in X-rays seems uncertain. Two large projects, NASA's Constellation-X and the European XEUS (X-ray Evolving Universe Spectrometer), have been merged, with additional aid from Japan, in the hope of increasing their chances of agency funding in the IXO (International

---

[5] Pulsars are thought to be rapidly rotating neutron stars, that emit regular pulses of radio waves at rates of up to one thousand pulses per second.

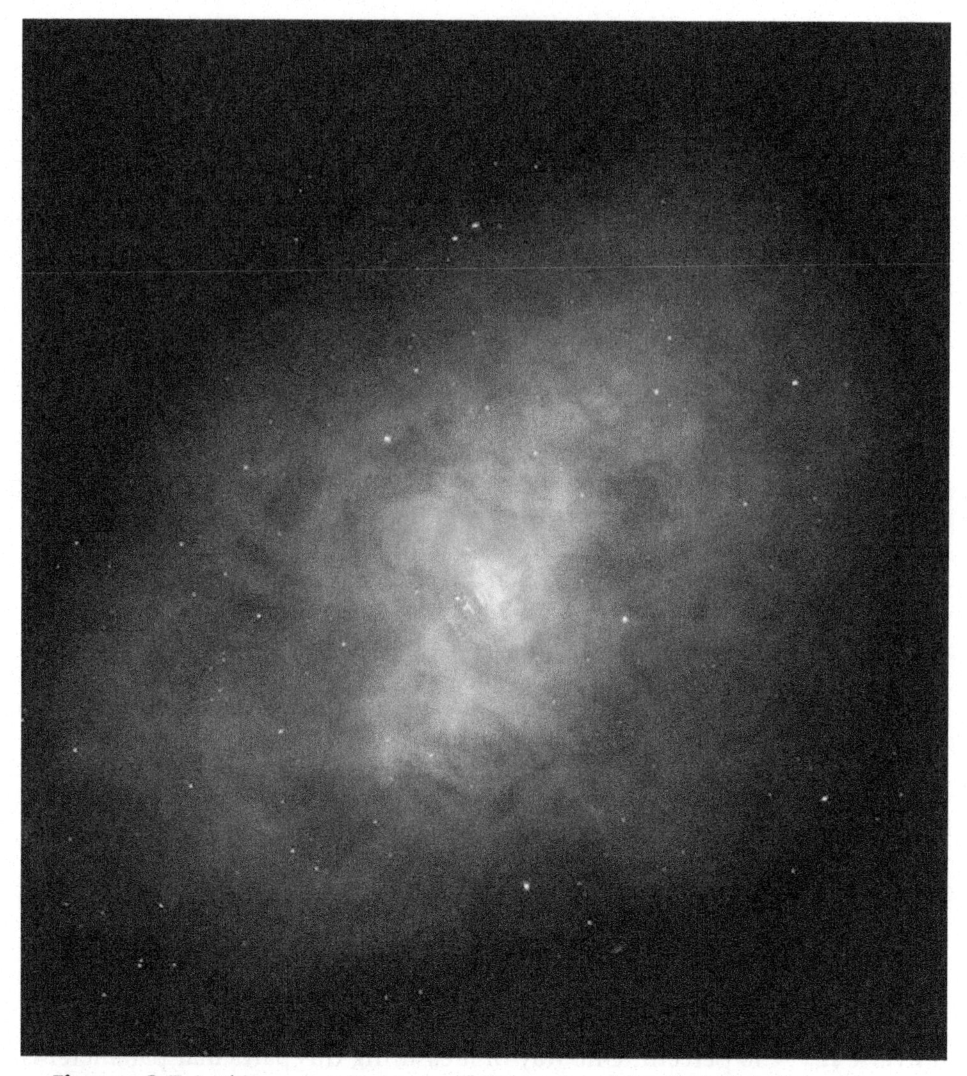

Figures 8.7 In this composite image of the center of the Crab Nebula, red represents radio emission, green represents visible emission, and blue represents X-ray emission. This nebula is the remnant of a supernova explosion observed in 1054. The dot at the very center is a pulsar. a neutron star, spinning 30 times per second, and is all that remains of the progenitor star. Note that X-rays mark the most energetic regions, i.e. the rotating disk of matter around the pulsar, while the (cooler) regions ejected during the explosion appear only in the radio and visible images. How does a neutron star, only 10-20 kilometers across, power the vast Crab Nebula? The expulsion of wisps of hot gas at high speeds appears to be at least part of the answer. Wisps like this likely result from tremendous electric voltages created by the rapidly rotating, magnetized, central neutron star. The hot plasma strikes existing gas, causing it glow in colors across the electromagnetic spectrum (J. Hester (ASU), CXC, HST, NRAO, NSF, NASA).

Figure 8.8 This composite image shows the massive galaxy cluster MACS J0717.5+3745 (MACS J0717, for short), where four separate galaxy clusters have been involved in a collision. Hot gas is shown in an image from NASA's Chandra X-ray Observatory, and galaxies are shown in an optical image from the Hubble Space Telescope. The hot gas is color-coded to show temperature, where the coolest gas is reddish purple, the hottest gas is blue, and the temperatures in between are purple. The repeated collisions in MACS J0717 are caused by a 13-million-light-year-long stream of galaxies, gas, and dark matter, known as a filament, pouring into a region already full of matter. A collision between the gas in two or more clusters causes the hot gas to slow down. However, the massive and compact galaxies do not slow down as much as the gas does, and so move ahead of it. Therefore, the speed and direction of each cluster's motion – perpendicular to the line of sight – can be estimated by studying the offset between the average position of the galaxies and the peak in the hot gas (NASA, ESA, CXC, C. Ma, H. Ebeling and E. Barrett (University of Hawaii/IfA), *et al.* and STScI).

Figure 8.9 An artist's impression of NASA's Chandra X-ray Observatory, which was launched and deployed by Space Shuttle Columbia on 23 July 1999. Chandra is designed to observe X-rays emanating from very high-energy regions of the universe. The observatory has three major parts: the X-ray telescope, whose mirrors focus X-rays from celestial objects; the science instruments which record the X-rays so that X-ray images can be produced and analyzed; and the spacecraft, which provides the environment necessary for the telescope and the instruments to work. Chandra's unusual orbit was achieved after deployment by a built-in propulsion system which boosted the observatory to a high Earth orbit. This elliptical orbit takes the spacecraft more than a third of the way to the Moon before returning to its closest approach to the Earth of 16,000 kilometers. The time to complete an orbit is 64 hours and 18 minutes. The spacecraft spends 85 percent of its orbit above the belts of charged particles that surround the Earth. Uninterrupted observations as long as 55 hours are possible and the overall percentage of useful observing time is much greater than for the low Earth orbit of a few hundred kilometers used by most satellites (NASA Marshall Space Flight Center and the Chandra X-ray Center of the Smithsonian Astrophysical Observatory).

X-ray Observatory) project. However, in spite of this, IXO does not show up in the recent NASA decadal priorities. The initial ESA XEUS Project was designed to be a hundred times more powerful than XMM-Newton. One particular feature is that it consists of two satellites, fifty meters apart in space, one bearing the mirror and the other the instruments.

Four flamboyant (microwave) decades

Since the fortuitous discovery[6] in 1965 by Penzias and Wilson of the cosmic microwave background radiation which turned out to be the fossil signal of the hot phases of the universe, many efforts, both experimental and theoretical, have been made to measure and interpret it to ever greater degrees of accuracy. Just as physicist George Gamow had predicted in an historic article, and Robert Dicke and Jim Peebles had independently forecast a little later, the spectacular isotropy of the photons detected and their perfect conformity to the spectrum of a black body indeed confirmed the cosmological origin of this radiation.[7]

Perhaps the most extraordinary aspect was the detection of arcminute deviations (of just a few micro-degrees in temperature) in this isotropy: the fossil traces of initial fluctuations which were to lead to the great structures of the cosmos. These momentous discoveries, which made the headlines even in the non-scientific press, were the result of decades of effort, relying largely upon considerable advances in detector technology for their accomplishment. The cosmic microwave background, it was revealed, is black-body radiation at a temperature of the order of 3 K. It is emitted in the millimetric domain, and extraordinarily sensitive instruments have had to be developed to explore this domain, which is heavily filtered by the Earth's atmosphere, itself an emitter in the same domain. There are however some 'windows' for this radiation if detectors are placed at high altitude. So work is proceeding at sites on isolated mountains and high plateaux (for example, in Hawaii and Chile, and at the South Pole); balloons and, ultimately, spacecraft are being launched.

---

[6] The story is now well known. Experimenting with Bell Laboratories' new microwave horn antenna in Crawford Hill, New Jersey, Arno Penzias and Robert Wilson detected parasitic 'noise', which was eventually found to be the result of cosmological radiation. In 1978 they won the Nobel Prize for their discovery.

[7] Dicke and Peebles, having predicted background radiation in the microwave domain, were working with David T. Wilkinson on the detection of microwaves. Having learned of Penzias' and Wilson's discovery, Dicke, Peebles and Wilkinson opted for simultaneous publication of the results.

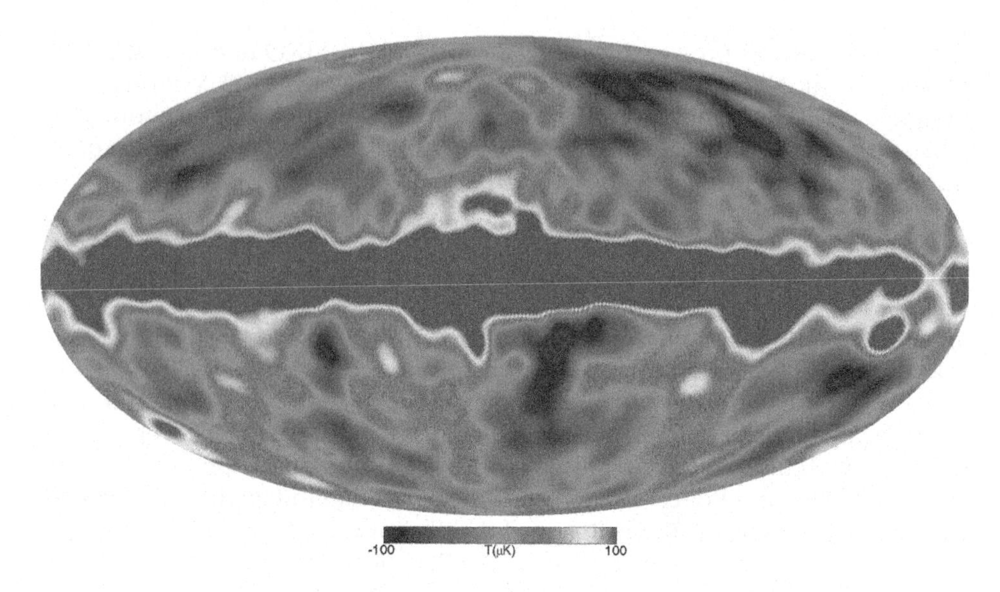

Figure 8.10 The all-sky image produced by the COBE Satellite. It is a low resolution image of the sky (7 degree resolution), but obvious cold and hot regions are apparent in the image. The large red band is the microwave emissions from our own galaxy. This image shows a temperature range of $\pm$ 100 microKelvin. It was processed through the same data pipe as the first year WMAP data (NASA/WMAP Science Team.)

It is not only the atmosphere that can interfere with measurements. There are many foreground celestial sources which also emit radiation at these wavelengths. The principal emitter is our Galaxy which appears as a highly visible band across the WMAP and COBE maps (Figure 8.10). Astronomers have succeeded in subtracting these parasitic signals by observing the offending sources at several wavelengths at which the cosmic background radiation is negligible. Obviously, this requires the availability of excellent data and the application of complex and delicate techniques.

As already noted, ground-based observers, to whom only certain 'windows' are available, find it difficult to eliminate these parasitic signals. However, one advantage of observing from Earth is that very large millimetric antennae can be deployed, which, with their very good angular resolution, can provide highly detailed temperature maps. On the other hand, observing from space means that not only is the detector above the atmosphere, but also, the whole sky can be mapped in the course of continuous long-term observations.

In 1989, NASA launched the COBE (COsmic Background Explorer), carrying three specialized instruments to study the cosmic microwave background: FIRAS (Far InfraRed Absolute Spectrometer) measured the cosmic microwave spectrum; DMR (Differential Microwave Radiometer) sought out differences in temperature as a function of position in the sky

Figure 8.11 NASA/NSBF personnel inflate the one million m^3 balloon which will carry the BOOMERANG telescope on its 10-day trip around the Antarctic continent. In order to make its exquisitely sensitive measurements, BOOMERANG is lifted above 99 percent of the atmosphere to an altitude of 35 kilometers (120,000 ft.). The continuous sunlight and stable air currents over Antarctica enable 10 to 20 day long stratospheric balloon flights. This launch was preceded by two months of assembly at McMurdo research station, and half a decade of development and construction by a international team of researchers (NASA/NSBF).

(the famous fluctuations); and DIRBE (Diffuse InfraRed Background Experiment) studied the diffuse infrared emission created by the totality of distant celestial bodies. FIRAS/COBE absolutely confirmed that the cosmic background radiation was indeed that of a black body at a temperature of 2.728 K. The DMR instrument, after two years of cumulative observations, was the first to detect temperature variations ($\sim$1 part in 100,000), interpreted as resulting from fluctuations in the density of the cosmological fluid.

The variations in temperature measured by COBE were on relatively large scales as a result of the limited angular resolution of the instrument, which could not detect the finest details sought. However, its discoveries set off a competition among observers for ever finer measurements, which would increase our understanding and definition of the cosmological model. Balloon-borne experiments such as BOOMERANG (Figure 8.11), MAXIMA

Figure 8.12 The detailed, all-sky picture of the infant universe created from seven years of WMAP data. The image reveals 13.7 billion year old temperature fluctuations (shown as color differences) that correspond to the seeds that grew to become the galaxies. The signal from our Galaxy was subtracted using the multi-frequency data. This image shows a temperature range of $\pm$ 200 microKelvin (NASA/WMAP Science Team).

and ARCHEOPS, in which various US and European teams collaborated, led to considerable advances. The Planck spacecraft, launched in May 2009 by ESA, will probably constitute the ultimate experiment, following NASA's WMAP satellite, which has so far (from the WMAP 7-year results released in January 2010 - Figure 8.12) provided the most accurate measurements ever obtained and has helped us determine values for cosmological parameters, such as the baryonic density Ω_b, as well as for Ω_T and Ω_m.

New windows on the cosmos

In the course of this book, we have seen that the 'missing' baryons probably take the form of 'warm' gas spread throughout the large structures of the universe. This gas, at a temperature of tens to hundreds of thousands of degrees, is too hot for neutral hydrogen to survive in it: collisions are too severe for the electrons of hydrogen atoms to remain attached to the central proton. However, the temperature is at the same time too low for the gas to emit 'hard' X-rays, as happens in galaxy clusters. It can be shown that one way of revealing the presence of this 'warm' intergalactic gas is to detect it in the far ultraviolet, through either absorption or emission. However, these kinds of measurements are hard to do: in the case of nearby objects, the wavelength is absorbed by the atmosphere and observations from space are needed. In the case of remote gas clouds, emissions are too faint to be detectable, and the absorption features are lost in the depths of the Lyman-alpha forest and are almost impossible to distinguish.

The most promising way ahead seems to lie with the detection from space of the emission lines of five-times-ionized oxygen, since, as already stated, this gas has been 'polluted' with heavy elements ejected by several generations of stars.

Certain projects have attempted to study such emissions. For example, the aim of FIREBALL (Faint Intergalactic-medium Redshifted Emission BALLoon), sent up in 2007, was to demonstrate the feasibility of such research. Its first flight led to the refinement of detection methods and instrumentation, but no measurements could be taken. Another flight occurred in 2009. The observations went normally, and three targets have been observed in good conditions. Data analysis is ongoing, but is a difficult task due to balloon movements and detector sensitivity.

The ISTOS (Imaging Spectroscopic Telescope for Origins Surveys) project, an American initiative with French participation, will involve a satellite designed to discover and map the hidden baryons of the universe. This project is not high on the agencies' lists of short-term priorities at present, but its scientific importance will certainly ensure that it emerges from obscurity, and its dedicated satellite will make long-term observations, increasing the sensitivity of detection and the areas of sky covered.

A further, big question remains unanswered: the exact date for the re-ionization. As we have seen, there are disparities between the results from SDSS, based on their observations of quasars, and those of WMAP based on CMB temperature fluctuations. Were there in fact several phases of re-ionization? One way of testing the hypothesis and dating the different phenomena is through direct observation of neutral hydrogen at its well-known 21-centimeter wavelength. However, since we are here dealing with redshifts between 6 and 20, state-of-the-art detectors, working in domains from about 1 to 10 meters, will be needed.

The call will be answered by radio telescopes such as LOFAR (LOw Frequency ARray) and SKA (Square Kilometer Array). The SKA project involves a radio telescope covering one square kilometer, and will be 100 times more sensitive than any current installation. Four countries (Argentina, Australia, China and South Africa) have been accepted as candidate locations for SKA, at relatively high-altitude sites offering very dry climates, since atmospheric water absorbs the radio wavelengths that the astronomers aim to study. The final site selection is expected in 2012. The SKA will consist of a network of antennae each 6 meters across (Figure 8.13), and signals from hundreds of antennae will be combined to create heightened sensitivity and resolving power: of course, the bigger the telescope, the better it can distinguish between closely-spaced objects. In this case, although the SKA's detecting area is not completely filled with antennae, its enormous size will enable it to 'split' objects only 0.01 seconds of arc apart. This resolving power, allowing

Figure 8.13 Artist's impression of the 15 meter × 12 meter offset Gregorian antennas within the central core of the Square Kilometer Array (SKA). Dish antennas like these will form a substantial part of the SKA; around 3,000 dishes are currently planned. Many aspects of the SKA dish design are without precedent, not only because of the large numbers of dishes required, but also because of the huge sensitivity that will result (SPDO/TDP/DRAO/Swinburne Astronomy Productions).

SKA to distinguish for example between two objects 15 meters apart on the Moon, is about 50 times greater than that of any current ground-based instrument, and rivals that of the big space telescopes. Another advantage of this system is that, unlike optical telescopes with their limited fields of view, the SKA will be able to cover a vast area of the sky: about seven degrees on a side, the size of 15 full Moons. Such capabilities will enable SKA to detect emissions from primordial gas clouds, which are not illuminated by stars, of as little as a billion solar masses (the mass of a dwarf galaxy) at the epoch of the re-ionization of the universe. We will be able to 'see' clouds of neutral gas ionized by the earliest quasars, and thereby gain insight into the way things were at that crucial epoch of the life of the universe. The LOFAR array (Figure 8.14) is a prototype project that is testing out the technologies and methods to be employed by the SKA.

So astrophysics pursues its search for ever greater knowledge of the secrets of the cosmos, delving ever deeper into its history. It has seen technical advances undreamed of only a short time ago. Telescopes tens of meters in diameter, and detectors with billions of pixels, functioning at all wavelengths, are being built, and supercomputers handling colossal amounts of data will analyze their observations, and simulate the universe in all its detail. In the evolution of its techniques, astrophysics now resembles particle physics, which has come to view the universe as the ultimate laboratory.

Figure 8.14 A view of the main core of the Low Frequency Array (LOFAR), located in the north-east of the Netherlands. LOFAR is a radio interferometer which observes the universe at low radio frequencies, close to the FM radio band, from 90 to 200 MHz. LOFAR is novel in its design because it is the first telescope which can look at the entire sky at the same time, unlike other telescopes which you have to point. In order to look in any given direction in the sky one uses a supercomputer to obtain information from a given direction. In addition to the many LOFAR stations in the Netherlands, others are located in France, Germany and one in the UK. One of the key LOFAR projects is to hunt for a signal from the epoch where the universe became re-ionized, most likely due to stars or quasars in the early universe emitting photons which ionized the diffuse gas in the intergalactic medium.. LOFAR will be able to probe this epoch by looking at the redshifted transition of the hyperfine structure in neutral hydrogen. This transition has a rest frequency of 1.4GHz, so if reionization happened at a redshift of around $z = 8$ then a radio interferometer would be able to map it (Netherlands Institute for Radio Astronomy (ASTRON)).

9 From telescopes to accelerators

"Looking at the sky through a telescope is an indiscretion." Victor Hugo

If Victor Hugo was correct, then astronomers are becoming ever more indiscreet as the size of their telescope mirrors continues to increase, to as much as several tens of meters in the case of certain current projects. Should they wish, however, to pursue their indiscretion along other avenues, the large-scale study of the universe is intimately linked with its comprehension on the atomic and subatomic scales. Have we not encountered, throughout the preceding chapters, the notion of particles, baryons, dark matter and dark energy, nuclear reactions... the very vocabulary of particle physics, leading us towards a better understanding of the object of our studies in cosmology – the universe in its totality?

Indeed, cosmology and fundamental physics are closely entwined. The fundamental interactions of the history of the cosmos and the constituents of the universe are well suited to scrutiny by particle physicists. Moreover, the progress of our understanding of the unification of these fundamental interactions will require experiments involving energies we cannot hope to harness, so the cosmos (and especially its earliest moments) must be our ultimate particle accelerator. Theoretical physics and particle physics may be the keys to our knowledge of the cosmos, but astrophysics and (especially observational) cosmology have contributed much to physics itself. Stellar physics has suggested that the oscillation of neutrinos explains their absence in neutrino detectors designed to intercept these particles emanating from the Sun; studies of primordial nucleosynthesis have anticipated what accelerators have confirmed, about the family number of neutrinos; and the dynamics of galaxies, clusters and the universe have revealed the existence of dark matter and dark energy.[1]

---

[1] To explain the acceleration in the expansion of the universe, it has been necessary to postulate the existence of a component of its energy/matter content with a 'repulsive' action; cosmologists call it dark energy (because it lends acceleration while not in itself being identified). The cosmological constant is a possible candidate for this energy.

Another standard – and its extensions

Like cosmology, particle physics has its own 'standard model'. It can take pride in its immense successes in describing the intricate realms of the atomic and the subatomic, and it is daily put to the test in the course of experiments carried out in the world's largest particle accelerators. According to this model (Figure 9.1), the fundamental constituents of ordinary matter are quarks (the most recently announced being the 'top' quark, discovered in 1995 using Fermilab's Tevatron accelerator) and electrons (members of the lepton family). Quarks and leptons are elementary particles known as fermions: composite fermions, such as protons and neutrons, are key building blocks of matter. Interactions manifest themselves through the exchange of 'carrier' or 'messenger' particles, called bosons. Unlike fermions, bosons do not obey the Pauli Exclusion Principle.[2] There are two kinds of bosons. One category acts as a vector or mediator particle for a given interaction (the photon for electromagnetism, gluons for the strong interaction, W and Z for the weak, and the (still hypothetical) graviton for quantum gravity). The other category includes the (still undiscovered) Higgs boson, supposed to be responsible for the mass of the other particles. One of the tasks of the Large Hadron Collider (LHC) is to discover the Higgs boson. There are four kinds of interaction. The strong nuclear interaction, mediated by gluons, binds quarks together to form protons and neutrons. These form nuclei, in spite of the electrical repulsion of the protons. The electromagnetic interaction, mediated by photons, binds the electrons to the nuclei. In this way, light can be emitted during electromagnetic exchanges within plasma, for example. The weak force, mediated by the W and Z bosons, is responsible for the phenomena of spontaneous radioactivity, involving the disintegration of nuclei. It plays an essential part in the nuclear reactions that take place in the cores of stars. Gravitational interaction is mediated by (hypothetical) gravitons, and is the predominant force on the cosmic scale.

Even though we now understand them so well, a fundamental question still remains to be asked about the origin and diversity of these interactions. Might they not in fact be the result of some unique interaction which would unify them under certain physical conditions? The unification of the weak and electromagnetic interactions is now established, earning the Nobel Prize in 1979 for Sheldon Glashow, Steven Weinberg and Abdus Salam, and the W and Z 'messenger' bosons involved were found at CERN in 1983. Extensions of the standard model, such as supersymmetry (known as SUSY and pronounced 'Susie'), predict that all the interactions will converge at high

---

[2] The Pauli Exclusion Principle states that no two fermions may occupy the same *quantum state* simultaneously. This principle helps to explain the electron shell structure of atoms.

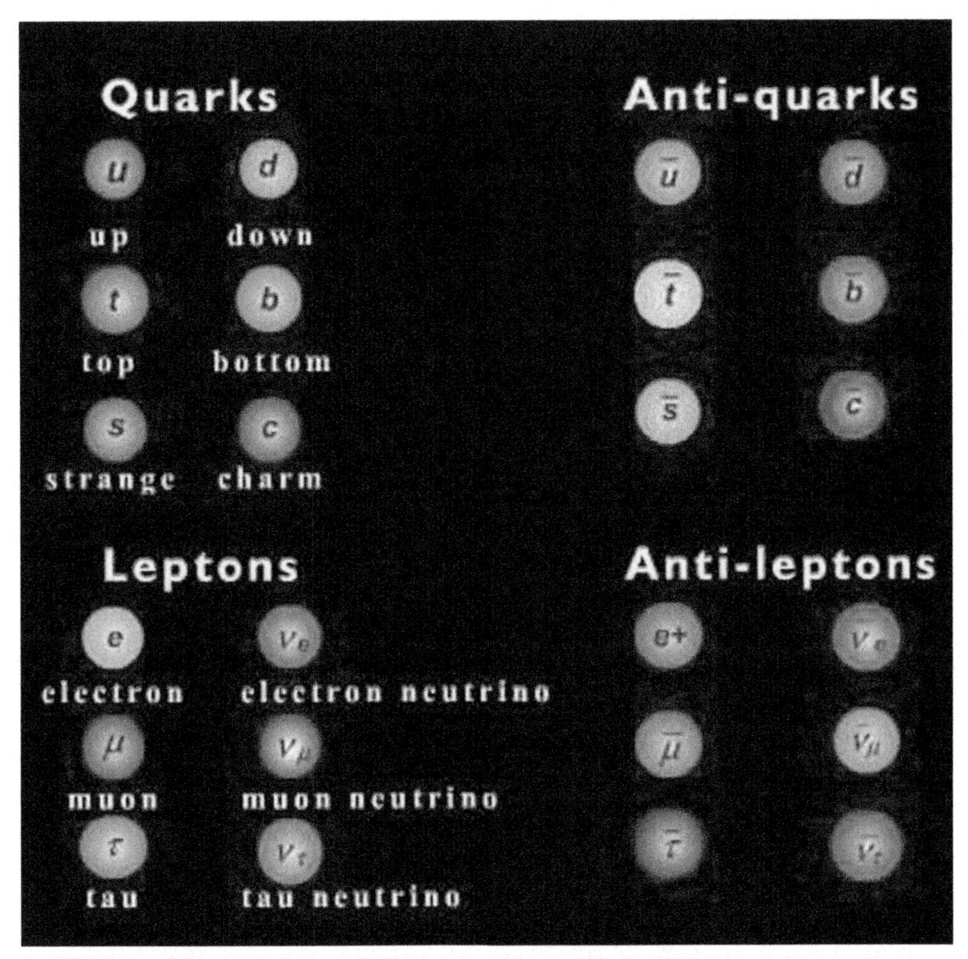

Figure 9.1 Elementary particles and their anti-particles. Quarks and leptons are elementary particles known as fermions: composite fermions, such as protons and neutrons, are the constituents of atomic nuclei and, with electrons, form ordinary (baryonic) matter. The neutrino, despite certain of its properties and the fact that its existence is recognized, is not a likely candidate for dark matter (IN2P3).

energies into one single interaction. However, these intriguing hypotheses have not yet been experimentally borne out in the case of the strong nuclear force and, *a fortiori*, gravity.

The intriguing hypothesis of supersymmetry has other direct and beneficial consequences for the cosmological model. One of these is connected with the question of anti-matter, briefly mentioned above. In the course of the thermal history of the universe, according to the standard model of particle physics, matter and anti-matter should have been produced in equal quantities, baryons and anti-baryons then annihilating each other almost totally. However, in the standard context, the prediction for the residual quantity of matter is less, by orders of magnitude, than astronomical observation suggests. Also, we should be observing radiation emitted at the edge of zones where matter and anti-matter meet, and this is not the case. The supersymmetry model, by letting go of certain restrictive hypotheses of the standard model, can apparently offer a natural explanation of the history of anti-matter.

Neutrinos as dark matter?

In Figure 9.1 we see that electrons, members of the family of leptons (particles that do not 'feel' the strong force), are accompanied by another category of particles: neutrinos. The existence of these particles, first proposed by Wolfgang Pauli in 1930 to preserve the conservation of energy in beta-decay reactions, was experimentally confirmed in 1956. Only in 1990 was it demonstrated that the total number of neutrino families is three. Like other particles, neutrinos were created during the hot phases of the primordial universe, aged about one second. Since that epoch, they have disconnected from all other matter and radiation and constitute (in the manner of the photons, 300,000 years on), a cosmic background of neutrinos bathing the whole cosmos. In our everyday environment, as well as the photons of the cosmic microwave background, there are about 150 'cosmological'[3] neutrinos per cubic centimeter, their energy greatly weakened by the fact of cosmic expansion. The temperature associated with this neutrino background is 1.9 K, not far removed from that of the photons of the CMB.

One thing in favor of neutrinos is that they belong to a recognized family of particles and are neutral, very abundant and almost never interact with other matter. So the Earth, continuously traversed by billions of neutrinos, 'stops' only very few of them. Their very abundance and the fact that they are neutral (and therefore emit no electromagnetic radiation) led cosmologists and physicists to believe, in the 1980s, that they had solved the problem of dark matter, having identified a family of particles in their accelerators. Indeed, while the standard particle model attributed zero mass to them, experiments at the time indicated that they were of non-zero mass. And that mass was such that the totality of neutrinos seemed to contribute enough to ensure a value of unity for the cosmological parameter Ω_{tot} – and no dark energy required! Alas, these estimates of neutrino mass later had to be revised downwards. The mass of the neutrino was still non-zero, but not enough for it to constitute the dark matter that dominates the dynamics of clusters and galaxies, and marshals the formation of large structures in the universe. If it isn't baryons, or neutrinos... we'll have to look elsewhere.

---

[3] Other, far more energetic neutrinos are also present, coming from bodies such as the Sun, supernovae, gamma-ray bursts, etc.

Direct and indirect research

So, although it possesses the expected attributes making it the prime candidate for dark matter, the neutrino has not quite been able to assume this lofty title. The problem remains that the standard model can put forward no other applicant. Fortunately for those astrophysicists casting about for candidates, most of the variants of supersymmetry theory predict the existence of a neutral and stable particle which might just fit the bill: the neutralino. However, this remains in the realms of theory.

The detection of this particle would at a stroke solve an astrophysical enigma, and open up new vistas in particle physics. The ideal situation would be a direct detection of the neutralino in a laboratory interaction experiment. We know though that such a detection would be extremely difficult to achieve, since the neutralino, like the neutrino, shows almost no interaction with ordinary matter in detectors, and the risk of parasitic effects is always present. Many experiments have been mounted during the last two decades, but the only really positive result has been that claimed by the DAMA team, working at the Gran Sasso laboratory in Italy, though scientific consensus is lacking.

Another, indirect, method consists in searching for possible products of the annihilation of dark-matter particles (these would be their anti-particles), emanating from areas of the sky in which they are concentrated, such as the centre of our Galaxy. These annihilation products may be gamma-ray photons, currently sought by the High Energy Stereoscopic System (HESS), an array of telescopes located in Namibia (Figure 9.2), and with which French researchers are associated. Another possible product, and putting in an unexpected reappearance here, is the neutrino. Neutrino telescopes are currently in operation, but they bear little resemblance to the traditional astronomers' telescopes. The Earth itself, capable of stopping the odd neutrino, is utilized, and such instruments scan the sky from the depths of our planet, sunk in the sea or buried in ice. The (rare) interaction between a neutrino (a product of the annihilation of dark matter) and the Earth creates a new particle known as a muon. The muon can be detected within a large, monitored volume of water or ice. What happens is that the muon, passing though these media, emits a luminous ray (Cherenkov radiation[4]) which

---

[4] Cherenkov radiation (also spelled Čerenkov) is electromagnetic radiation emitted when a charged particle (such as an electron) traveling at relativistic speeds moves through a medium at a speed greater than the velocity of light in that medium. When that happens, the particle emits radiation in the form of a 'shock wave', widely known as Cherenkov radiation. The characteristic blue glow of nuclear reactors is due to Cherenkov radiation. It is named after the Russian scientist Pavel Alekseyevich Cherenkov.

Figure 9.2 Two of the four HESS telescopes located in Namibia, near the Gamsberg mountains, an area well known for its excellent optical quality. HESS is a system of imaging atmospheric Cherenkov telescopes that investigates cosmic gamma-rays in the 100 GeV to 100 TeV energy range. The name HESS stands for High Energy Stereoscopic System, and is also intended to pay homage to Victor Hess, who received the Nobel Prize in Physics in 1936 for his discovery of cosmic radiation. The instrument allows scientists to explore gamma-ray sources with intensities at a level of a few thousandths of the flux of the Crab nebula (the brightest steady source of gamma-rays in the sky). All four of the HESS telescopes were operational by the end of 2003, and were officially inaugurated on 28 September 2004 (HESS Collaboration).

propagates in a cone (Figure 9.3). The detection of this radiation by multiple detectors is analyzed and its geometry reveals characteristics of the muon and (indirectly) those of the incident neutrino.

Two projects, ANTARES (Astronomy with a Neutrino Telescope and Abyss environmental RESearch project) and AMANDA (Antarctic Muon And Neutrino Detector Array), the first residing 2.5 km under the Mediterranean

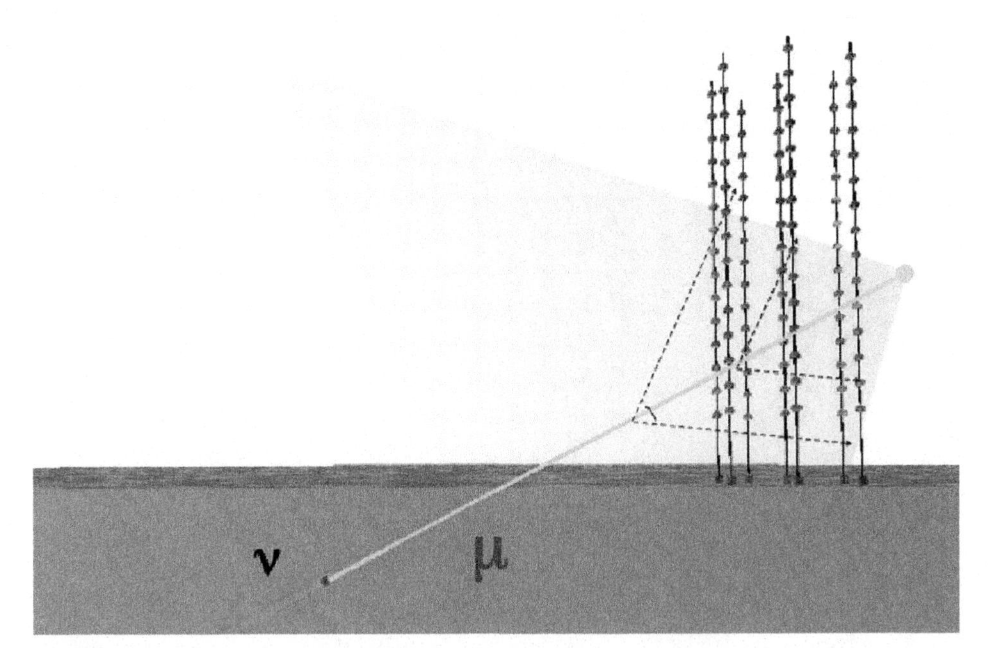

Figure 9.3 The principle of the neutrino telescope. Detection of the Cherenkov radiation emitted in a cone by muons created by the interaction of a neutrino, possibly a product of the annihilation of dark matter.

Sea off the coast of Toulon, France (Figure 9.4), and the second buried almost 2 km below the ice of the Amundsen-Scott South Pole Station, hunt for these elusive particles that may be messengers from one of the main constituents of the universe. In 2005, after nine years of operation, AMANDA officially became part of its successor project, the IceCube Neutrino Observatory (Figure 9.5).

What dark energy?

Not only does baryonic matter represent but a feeble component of the total matter of the cosmos, but its energy content, it has to be said, is dominated by dark energy – a strange kind of energy indeed, with its negative pressure that acts against gravity and accelerates the expansion of the universe. Evidence for its existence comes initially from observations of remote supernovae which indicate that the expansion of the universe is accelerating.

Einstein's theory of relativity accords quite easily, however, with this somewhat surprising finding, if we introduce the cosmological constant or infer the existence of a new form of energy with the same properties as those of the vacuum energy of quantum mechanics. It is not too difficult to

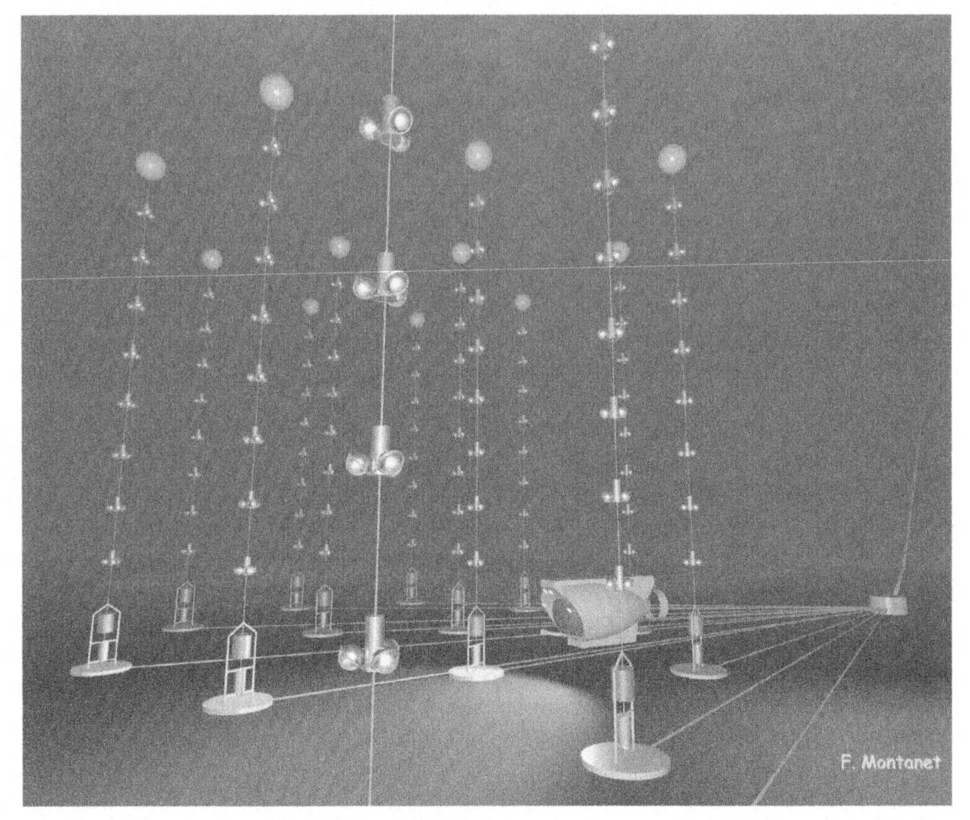

Figure 9.4 The ANTARES neutrino telescope. The detector consists of a network of 1,000 photomultipliers sensitive to the Cherenkov radiation created by muons (the product of the interaction of a neutrino with the environment of the detector), which travel faster than the speed of light in this medium. The photomultipliers are arranged in twelve vertical strings 350 meters high and about 70 meters apart, spread over an area of about 0.1 km^2. ANTARES works at a depth of 2,500 meters, 40 km from Toulon in the Mediterranean Sea (F. Montanet/ANTARES).

imagine the existence of new dark-matter particles, which would in the final analysis be (distant) cousins of the neutrinos, sharing their elusive properties. The same kind of leap of the imagination is not so easy in the case of dark energy, given its properties. Perhaps we are less taken aback when we recall that a similar (or the same?) kind of energy was involved in the inflationary phase of the primordial universe, whose description requires a marriage (as yet unconsummated) between general relativity and quantum mechanics. In the absence of an exact theory, astrophysicists and physicists are collaborating to try and find the best description of this still hypothetical dark energy, with its profound consequences for our fundamental knowledge of the universe and its laws. The elucidation of its nature is a complex problem,

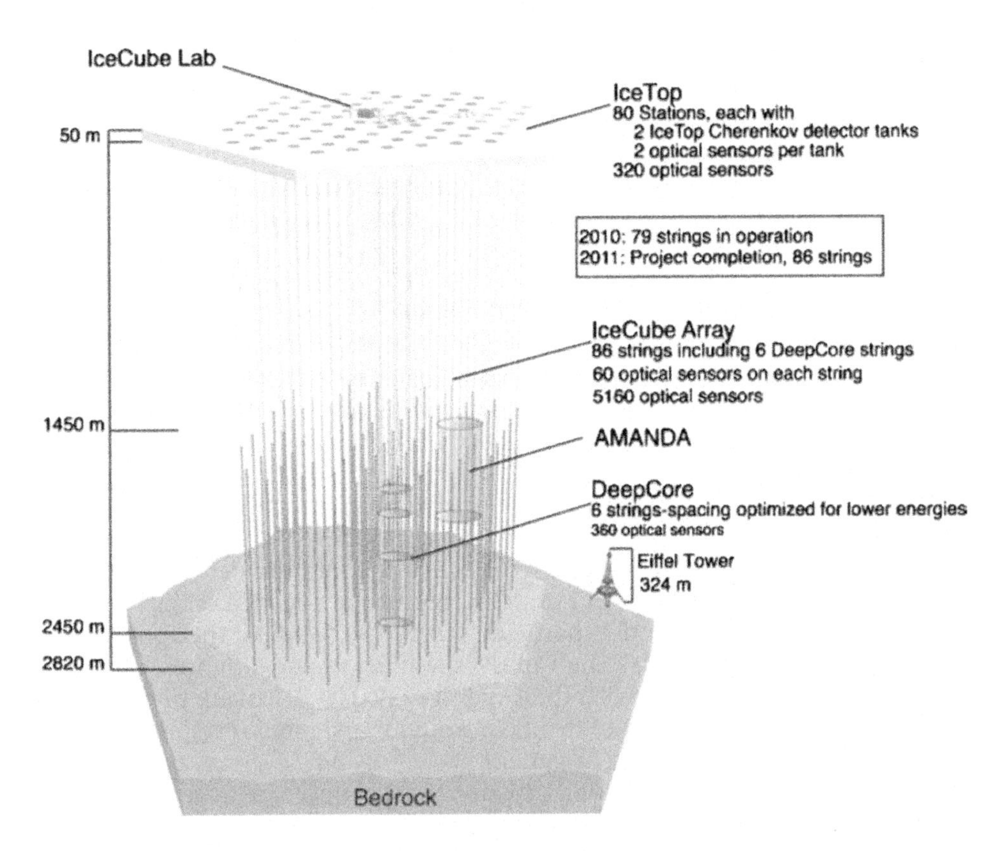

Figure 9.5 The IceCube Neutrino Observatory (or simply IceCube) is a neutrino telescope constructed at the Amundsen-Scott South Pole Station in Antarctica. Similar to its predecessor, the Antarctic Muon And Neutrino Detector Array (AMANDA), IceCube contains thousands of spherical optical sensors called Digital Optical Modules (DOMs), each with a photomultiplier tube (PMT) and a single board data acquisition computer which sends digital data to the counting house on the surface above the array. IceCube was completed on 18 December 2010 (IceCube Science Team, Francis Halzen, Department of Physics, University of Wisconsin).

especially the determination of its possible evolution through time. So fundamental physics is taking up the tools of astrophysics, blending results from all possible cosmic markers: supernovae, gravitational lenses, measurements of the CMB, counting clusters of galaxies... The coming decade should see many observational programs, both ground-based and in space, dedicated to dark energy and pursued jointly by researchers using both telescopes and accelerators. Examples are WFIRST (NASA) and Euclid (ESA).

Towards the Big Bang?

There are other important aspects of subatomic physics that remain poorly understood. Physicists would like to know, for example, why do all particles not have the same mass – and why should they have mass at all? This is one of the tasks of the Large Hadron Collider (LHC), which accelerated its first beam of particles in September 2008. This impressive machine is the latest in a line of accelerators descended from a $\sim$100 MeV cyclotron installed at Berkeley just after the Second World War. Decades later, the essential discoveries (the electron, proton, neutron, positron, etc...) having been made, the golden age of particle physics was ushered in by such machines. The LHC replaced the Large Electron-Positron Collider (LEP), which operated from 1989 until 2000, and achieved energies of 200 GeV (1,000 times greater than those of the earliest machines).

The LEP provided evidence of the electroweak interaction and measured the masses of the associated Z and W bosons. As far as the actual mass of particles is concerned, an extension of the standard model suggests the existence of an omnipresent 'Higgs field', with an associated 'Higgs boson'. The interaction of the other particles with this field is thought to be responsible for the genesis of their mass: the most massive/lightest particles interacting strongly/weakly with the Higgs boson. The discovery of the Higgs boson, and the measurement of its mass, would mark a major step forward in this discipline and for physics in general.

Finally, as we write the thermal history of the universe from its earliest moments, we require the intervention of physics capable of describing situations corresponding to ever higher temperatures: we have indeed seen that, as we go further back in cosmic time, the temperature increases very

Figure 9.6 The Large Hadron Collider (LHC) accelerates two beams of atomic particles in opposite directions around the LHC ring, which lies in a tunnel 27 km in circumference, beneath the Franco-Swiss frontier. When the particle beams reach their maximum speed, the LHC allows them to 'collide' at four points on their circular journey. (Top) This diagram shows the locations of the four main experiments (ALICE, ATLAS, CMS and LHCb) that will take place at the LHC. Located between 50 and 150 meters underground, huge caverns have been excavated to house the giant detectors. The SPS (Super Proton Synchrotron), the final link in the pre-acceleration chain, and its connection tunnels to the LHC are also shown. (Bottom) View of the LHC cryo-magnets inside the 27-km long LHC tunnel. It is vital that each magnet is placed exactly where it has been designed so that the path of the particle beam is precisely controlled (CERN, Geneva).

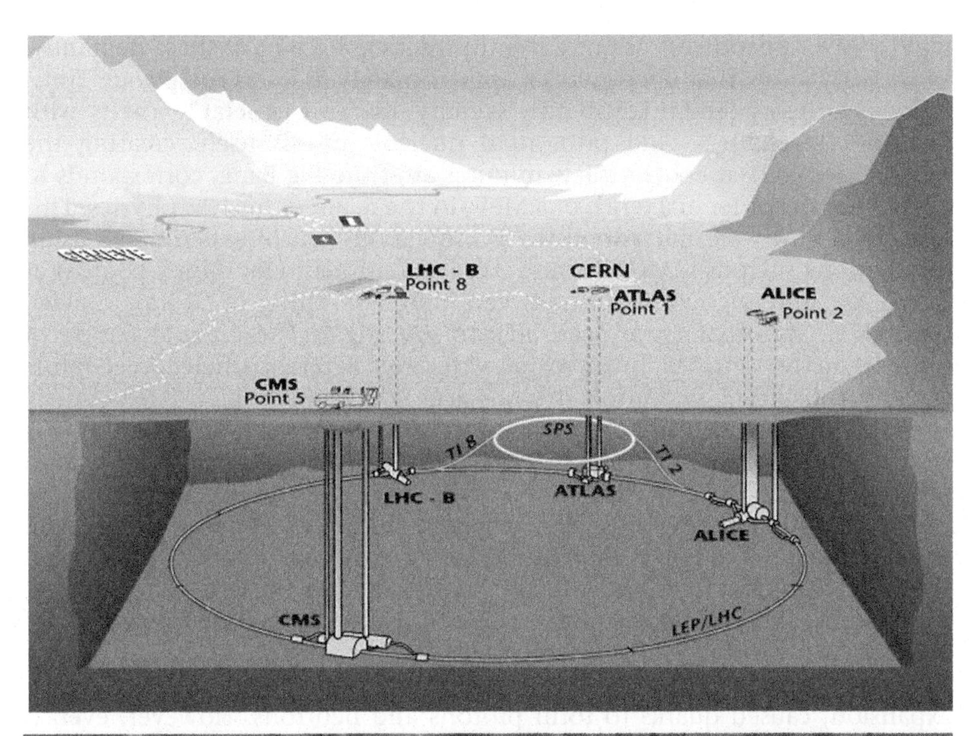
LHC - B
Point 8
CERN
ATLAS
Point 1
ALICE
Point 2
CMS
Point 5
SPS
TI 8
TI 2
LHC - B
ATLAS
ALICE
CMS
LEP/LHC

rapidly (see Appendices). Is our current physics able to meet these demands? We already know that if we go back approximately as far as the Planck Time, we need a theory (and it is still only sketchy) merging general relativity with quantum mechanics. And primordial nucleosynthesis itself, creating the lightest elements at epochs a few minutes after the Big Bang, corresponds to energies of the order of a tenth of a MeV, in the realm of nuclear physics. This domain is now well understood thanks to various sets of experiments using installations such as GANIL (Grand Accélérateur National d'Ions Lourds) at Caen. By studying collisions between different kinds of atomic nuclei, researchers at GANIL have been able to investigate the integral nature of nuclear matter and its interactions. Are ever larger particle accelerators actually capable of recreating the physical conditions that existed shortly after the Big Bang, during the first second of the universe, an epoch during which these fundamental interactions, and elementary particles, dominated the stage?

The LHC, now the world's most powerful particle accelerator (Figure 9.6), should attain, through collisions of heavy particles, temperatures about 100,000 times that at the centre of the Sun, i.e. about a thousand billion degrees. It will therefore become possible to study in detail that phase of the universe ($t \sim 10^{-11}$ s) during which the cosmos consisted of a hot, dense soup: the quark-gluon plasma that existed before cooling, driven by expansion, caused quarks to form protons and neutrons. However, even if we look back to 10^{-11} seconds, our search will be far from over, because the unification of all interactions must occur at an epoch around 10^{-39} seconds (10^{16} GeV), and finally the Planck era, at 10^{-43} seconds (10^{19} GeV).

So, not only cosmologists can play with as many kinds of universe as they want on their computers, but particle physicists are also able to reproduce primordial cosmic fluids in their laboratories!

Lemaître, Einstein, Gamow and others would have certainly like that ....

What's up with particle physics?

As we have already seen, and for still not understood reasons, anti-matter is almost absent from the universe. One way to resolve this mystery would be to handle that anti-matter directly! Although at first sight this seems incredible, anti-matter is created daily in particle accelerators. From the late 1990s, the ATHENA experiment at CERN produced low-energy anti-hydrogen atoms for comparison with hydrogen atoms. In September 2002, ATHENA announced the first controlled production of a large number of anti-hydrogen atoms at low energies and the direct observation of their annihilation (Figure 9.7). The ATHENA experiment ended in November 2004 and an experiment called ALPHA has since been set up to continue the research begun by its predecessor. In late 2010, physicists working on the ALPHA experiment made an important step not only in producing anti-matter, but also in

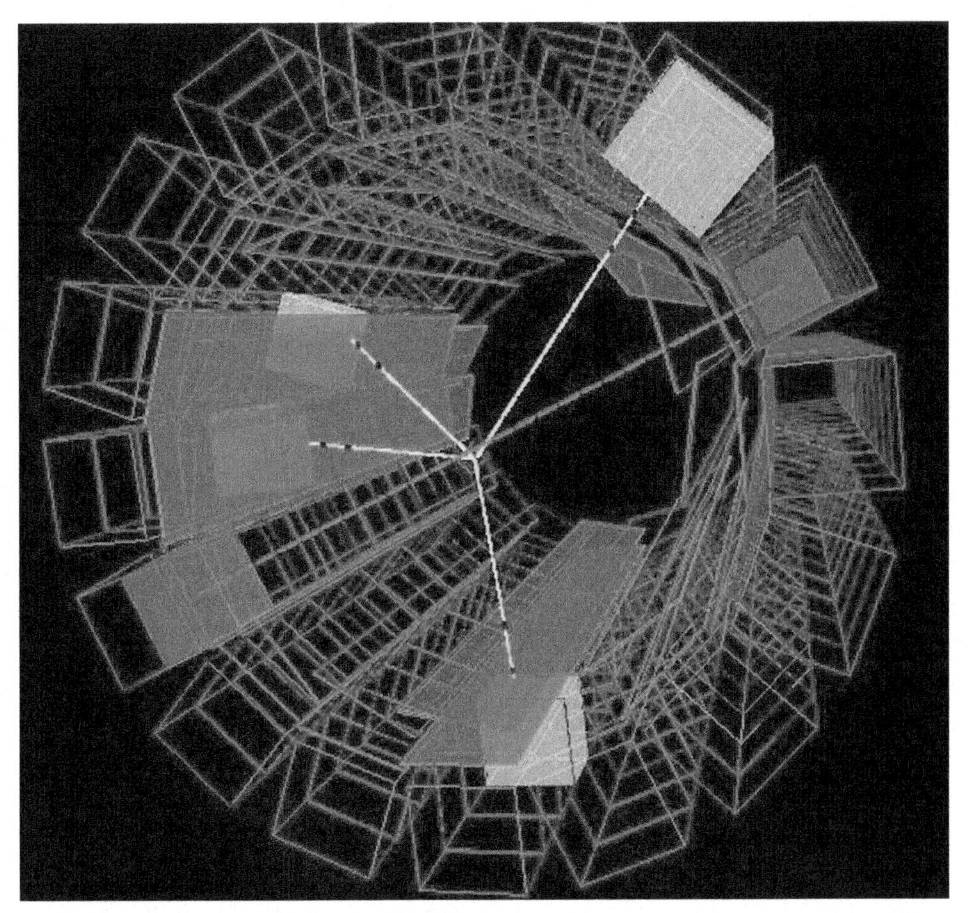

Figure 9.7 This is an image of an actual matter-antimatter annihilation due to an atom of antihydrogen in the ATHENA experiment, located on the Antiproton Decelerator (AD) at CERN since 2001. The antiproton produces four charged pions (yellow) whose positions are given by silicon microstrips (pink) before depositing energy in CsI crystals (yellow cubes). The positron also annihilates to produce back-to-back gamma rays (red) (CERN, Geneva).

succeeding to build and trap anti-atoms of hydrogen (the most simple to build), preventing them from annihilation with ordinary matter by huge magnetic fields (even if anti-atoms are also globally neutral). They trapped 38 anti-atoms of hydrogen for about one fifth of a second, long enough to study them in detail; previous attempts at CERN, beginning in 1995, had led to lifetimes of only 40 billionths of a second. Then, in May 2011, the ALPHA team described how it trapped 309 anti-hydrogen atoms for $\sim$1000 seconds. This boost in both number and trapping time should lead to important insights into the nature of antimatter versus matter.

Also at CERN, another experiment named ASACUSA aims to learn more about fundamental differences in the behavior of matter and anti-matter. Instead of directly comparing atoms with their corresponding anti-atoms (as in the ALPHA experiments), ASACUSA creates hybrid atoms such as 'anti-protonic helium'. Helium has the second simplest atomic structure after hydrogen; it contains two electrons orbiting a central nucleus. An 'anti-protonic helium' atom is made by replacing one of these orbiting electrons with an anti-proton. The process of creating these hybrid atoms is easier than making anti-hydrogen atoms, and they can be kept for longer. The ASACUSA team uses the Anti-matter Decelerator at CERN to send a beam of anti-protons into cold helium gas. Most of the anti-protons quickly annihilate with ordinary matter in the surroundings, but a tiny proportion combines with the helium to form hybrid atoms that contain both matter and anti-matter. Using laser beams to excite the atoms, ASACUSA can measure the mass of the anti-proton to an unprecedented level of accuracy for comparison with the proton.

Moreover, in the near future the AEGIS experiment at CERN will attempt to observe how anti-matter responds to gravity and if it differs from ordinary matter in some respect. *A priori*, within the present framework of theoretical physics, no difference in the respective behavior is foreseen. But, who knows? Progress in science is generally a consequence of unexpected results, so the experiment should be done. A hope is that all of these new investigations will finally help to solve one of the most irritating cosmic enigmas: the apparent absence of anti-matter in our Universe.

Figure 9.8 Two of the four huge underground caverns that have been excavated at the four collision points located around the LHC ring. Detectors placed around the collision points are able to follow the billions of collisions and new particles produced every second and identify the distinctive behavior of interesting new particles from among the many thousands that are of little interest. The dimensions and extreme complexity of these experiments are suggested by these images. (Top) In the ALICE experiment, the LHC will collide lead ions to recreate the conditions just after the Big Bang under laboratory conditions. The data obtained will allow physicists to study a state of matter known as quark-gluon plasma, which is believed to have existed soon after the Big Bang. The ALICE detector is 26 meters long, 16 meters high, and 16 meters wide, and weighs 10,000 tons. (Bottom) ATLAS is one of two general-purpose detectors at the LHC. It will investigate a wide range of physics, including the search for the Higgs boson, extra dimensions, and particles that could make up dark matter. ATLAS will record sets of measurements on the particles created in collisions – their paths, energies, and their identities. The ATLAS detector is the largest volume particle detector ever constructed. It is 46 meters long, 25 meters high and 25 meters wide, and weighs 7,000 tons (CERN, Geneva).

After a brilliant start in early September 2008, immediately followed by a terrible failure (damaging both the instruments and the tunnel itself),[5] the LHC, reborn anew like the Pheonix, recently sent some good news about hot topics concerning cosmology and particle physics. After such promising advances in the anti-matter sector of particle physics, involving light particles, scientists at CERN using the LHC turned towards the use of heavy lead ions – lead atoms stripped of electrons – with the ALICE, ATLAS and CMS detectors towards the end of 2010 (Figure 9.8). Their aim was to realize one of the most crazy dreams of cosmologists: traveling back in time towards the Big Bang and probing *in the laboratory* the primordial cosmic soup (Figure 9.9).

With the lead ions used in the LHC beam collisions and the energy domain in play,[6] protons freed from the heavy nuclei are broken into their basic constituents: quarks and gluons which then form a plasma. In fact, this quark-gluon plasma is the form the cosmic soup is supposed to have exhibited when the universe was only 10^{-11} seconds old, so the physicist's dream starts becoming reality! But, to their surprise, and contrary to what was expected (but suggested by a previous experiment at the Relativistic Heavy Ion Collider at Brookhaven National Laboratory), this cosmic soup appears in the form of a liquid rather than a gas. Physicists now have to strengthen their efforts to study the detailed properties of this new kind of liquid and tell us what really happens at this epoch.

After many vicissitudes, and having already been tested aboard the Space Shuttle in 1998, the Alpha Magnetic Spectrometer (AMS-02), a particle physics experiment module, was launched in May 2011 aboard the penultimate flight of the US Space Shuttle. It was installed on the International Space Station (ISS) and will be operated for the lifetime of ISS (Figure 9.10). Enthusiastically promoted by the Nobel Prizewinner in Physics, particle physicist Samuel C.C. Ting, AMS is a particle detector that will track incoming charged particles such

---

[5] The incident on 19 September 2008 caused a tonne of liquid helium to leak out into the 27 km-long, ring-shaped tunnel that houses the LHC. The liquid helium is used to chill the 1,232 dipole 'bending' magnets arranged end-to-end in the tunnel to a temperature of 1.9 kelvin (–271°C). This makes the magnets superconducting, enabling them to produce the large magnetic fields required to steer the particle beams while at the same time consuming relatively little power. The problem caused around 100 of the LHC's super-cooled magnets to heat up by as much as 100 degrees, so their superconducting properties were lost. The most likely cause of the equipment failure was a faulty electrical connection between two of the accelerator's magnets. This connection melted during testing of the machine, causing a major leak of the super-cooled liquid helium.

[6]

[7] Once circulating beams had been established they could be accelerated to the full energy of 287 TeV per beam. This energy is much higher than for proton beams because lead ions contain 82 protons.

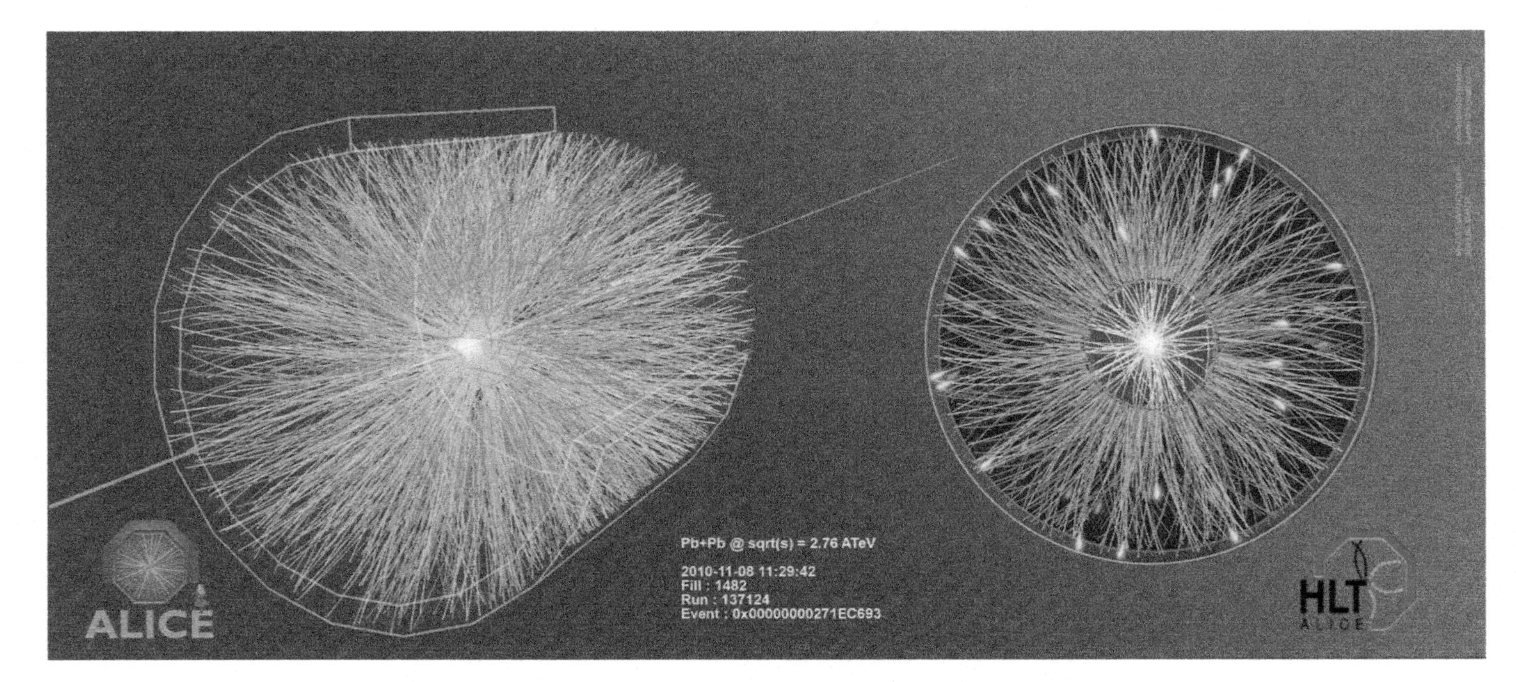

Figure 9.9 Tracks from the first heavy ions collisions in ALICE. On 8 November 2010, LHC scientists declared for the first time that they were running stable beams of heavy ions. Here we see the first events resulting from lead ion-lead ion collisions at a centre-of-mass energy of 2.76 TeV per nucleon pair. This is an online reconstructed event from the High Level Trigger (HLT), showing tracks from the Inner Tracking System (ITS) and the Time Projection Chamber (TPC) of ALICE (CERN, Geneva, ALICE Collaboration).

Figure 9.10 Two artists' impressions of the Alpha Magnetic Spectrometer (AMS-02), a particle physics experiment module, that was launched in May 2011 by the US Space Shuttle Endeavour, and installed on the International Space Station (ISS). AMS is a particle detector that will track incoming charged particles such as protons, electrons and atomic nuclei that constantly bombard our planet. (Top) A wide-angle view of the entire ISS showing the location of AMS-02 which is mounted on top of the Integrated Truss Structure, on USS-02, the zenith side of the S3-element of the truss. (Bottom) Close up view of AMS-02 *in situ*. mounted to the ISS S3 Upper Inboard Payload Attach Site (NASA).

as protons, electrons and atomic nuclei that constantly bombard our planet. By studying the flux of these cosmic rays with very high precision, AMS will have the sensitivity to identify a single anti-nucleus among a billion other particles. Although primarily designed to study cosmic rays, AMS will also help researchers study the formation of the Universe and search for elusive dark matter and cosmic anti-matter.

A cryogenic, superconducting magnet was originally developed for the AMS-02, but problems with the magnet's cooling system contributed to a decision to abandon the cryogenic system in favor of a previously developed, but less capable, permanent magnet system. Although this non-superconducting magnet has a weaker field strength, its operational lifetime in orbit is expected to be 10 to 18 years compared with only 3 years for the superconducting version. This additional data gathering time has been deemed more important than higher experiment sensitivity, Whether the ISS will operate long enough for AMS to take full advantage of its likely extended lifetime is unclear.

So, for cosmology and fundamental physics, there still exist many areas of confrontation and convergence in the quest for a coherent vision, acceptable to those who seek it, of both the microcosm and the macrocosm, of ordinary baryons and the most exorbitant energies.

Appendices

Powers of ten

Powers	Symbol	Values
10^{-12}	pico (p)	0.000000000001
10^{-9}	nano (n)	0.000000001
10^{-6}	micro (μ)	0.000001
10^{-3}	milli (m)	0.001
1		
10^{3}	kilo (k)	1 000
10^{6}	mega (M)	1 000 000
10^{9}	giga (G)	1 000 000 000
10^{12}	tera (T)	1 000 000 000 000
10^{15}	peta (P)	1 000 000 000 000 000

Units in high-energy physics

High-energy physicists use units of time, dimension, mass and temperature which are expressed as a function of the basic unit of energy: the electron volt (eV). This is the energy of one electron in an electrical field of one volt. Our usual system of notation would express it thus:

$$1 \text{ eV} = 1.6 \cdot 10^{-19} \text{ Joules}$$

It is therefore, as a tiny unit, well suited to describe elementary particles.

This system of units is obtained by ascribing the value of unity to the fundamental constants of physics:

$$h = c = k = 1$$

So we arrive at the equivalence:

	Temperature	*Mass*
1 eV →	11 600 K	$1.78 \cdot 10^{-30}$ kg

The Planck era

In the context of quantum physics, which deals with the atomic and subatomic realms, the energy of a system and exchanges of energy are 'quantified'.

The minimum quantum is set by Planck's Constant:

$$h = 6.626 \cdot 10^{-34}\ \mathrm{J \cdot s}$$

We also define the Reduced Planck Constant:

$$\hbar = h/2\pi = 1.054 \cdot 10^{-34}\ \mathrm{J \cdot s}$$

Gravitation is expressed, in the classic manner, by Newton's Laws, where we find the universal gravitational constant, also called Newton's Constant:

$$G = 6.673 \cdot 10^{-11}\ \mathrm{m^3/kg/s^2}$$

Its generalization occurs within the context of General Relativity, which involves the speed of light (c), defined as a universal constant:

$$c = 299{,}792{,}458\ \mathrm{m/s}$$

It is possible, with these three fundamental constants, to construct various quantities such as the Planck length, time, mass or temperature.

$$L_{Planck} = (\hbar\, G/c^3)^{1/2} \sim 10^{-35}\ \mathrm{m}$$
$$t_{Planck} = (\hbar\, G/c^5)^{1/2} \sim 10^{-43}\ \mathrm{s}$$
$$M_{Planck} = (\hbar\, c/G)^{1/2} \sim 10^{-8}\ \mathrm{kg}$$
$$T_{Planck}/k = (c^5 \cdot \hbar/G^2) \sim 10^{32}\ \mathrm{K}$$

Thermal history of the universe in brief

		Time/redshift $t \sim 1/5 \cdot T^2$ (MeV)	Temperature/energy
Planck era	All interactions unified	$\sim 10^{-43}$ s/10^{32}	$\sim 10^{32}$ K/10^{19} GeV
First breaking of symmetry	Gravity becomes distinct	$\sim 10^{-36}$ s/10^{28}	$\sim 10^{28}$ K/10^{15} GeV
Second breaking of symmetry	Strong and electroweak interactions become distinct	$\sim 10^{-32}$ s	$\sim 10^{27}$ K
Third breaking of symmetry	Separation of weak and electromagnetic interactions	$\sim 10^{-12}$ s/10^{15}	$\sim 3 \cdot 10^{15}$ K/250 GeV
Baryogenesis	Quarks confined within protons and neutrons	$\sim 10^{-6}$ s	$\sim 10^{13}$ K/1 GeV
Nucleosynthesis	Mainly H and He	~300 s–35 min	10^9 K/1 MeV
Matter dominates	(Dark) matter dominant	~60 000 years	10 000 K/1 eV
Recombination	Electrons and protons form atoms	300 000 years	~3 000 K
CMB	The universe becomes transparent	300 000 years/1100	~3 000 K
First galaxies		$5 \cdot 10^8$ years?/~15?	~10 K
Present day	Life	$14 \cdot 10^9$ years?/0	~3 K/0.0002 eV

Glossary

Accretion The accumulation of matter as a result of gravitational interaction. Black holes, with their strong gravity, attract material from their surroundings, forming an accretion disk.

Angström (Å) Unit of measurement used in atomic physics for wavelengths of electromagnetic radiation and atomic dimensions, with a value of 10^{-10} meter. The unit is named after the famous Swedish physicist, one of the inventors of spectroscopy. The official SI unit is the nanometer (= 10 Å).

Annihilation The transformation of a particle and an anti-particle into a shower of other particles, principally gamma-ray photons.

Anti-matter Anti-particles have the same mass as their 'mirror' particles, but they are of opposite quantum number, especially as regards their electric charge. When a particle of matter meets its anti-particle, they mutually annihilate, releasing all their energy (i.e. the sum of their kinetic energy and the energy corresponding to their mass).

Atom From the Greek *άτομος* (*atomos*, indivisible). Elementary constituent of matter, in fact composed of various fundamental particles (electrons, neutrons, protons). Nowadays, we know that atoms are certainly not 'indivisible'.

Baryon/baryonic From the Greek *βαρύς* (*baros*, heavy). A baryon is a hadron composed of three quarks. The most well known baryons are the protons and neutrons which make up the majority of the mass of the visible matter in the universe, 'Baryonic matter' is matter made of atoms, i.e. protons and neutrons (so-called 'ordinary' matter).

Black body A totally absorbent heated body. A good way to visualize this is by imagining an oven equipped with an opening to allow observation of the radiation in its interior. The hotter the body becomes, the shorter the wavelengths emitted as radiation ('white heat'), and vice versa. This wavelength λ (in cm) is related to the temperature T (in degrees K) by Wien's Law: $\lambda T = 0.29$. The 3 K cosmic microwave background emits black-body radiation in the millimetric waveband. The Sun, whose photosphere behaves as a black body at 5,300 K, therefore emits radiation at around 550 nm, which we see as yellow.

Boson Elementary particle. Unlike fermions, bosons do not obey the Pauli exclusion principle. There are two kinds of bosons. One category acts as a vector or mediator particle for a given interaction (the photon for electromagnetism, gluons for the strong interaction, W and Z for the weak, and the (still hypothetical) graviton for quantum gravity). The other category includes the (still undiscovered) Higgs boson, supposed to be responsible for the mass of the other particles. One of the tasks of the Large Hadron Collider (LHC) is to discover the Higgs boson.

Bremsstrahlung or free-free emission All charged particles deflected by their passage near another charged particle emit this continuous radiation.

Brown dwarfs See Dark Baryons

Classification of galaxies Galaxies can be classed into three main categories: spirals, disk-shaped systems with a central spherical region surrounded by rotating arms; ellipticals, shaped like rugby balls or soccer balls; and irregulars, of indeterminate shape. Most of the galaxies of the early universe were irregulars, and more regularly shaped systems formed later.

CMB The cosmic microwave background, the 3 K 'fossil' radiation filling the universe, discovered by Penzias and Wilson.

Cosmological Constant (Λ) A term introduced by Einstein into his cosmological equations, which enabled him to find solutions based on a static universe. The discovery of the recession of the galaxies by Hubble, and of cosmic expansion, cast doubt upon this constant (Λ). Cast aside for decades, it was resurrected as a possible cause of the acceleration of cosmic expansion. It has thus become a candidate for the famed dark energy responsible for this acceleration.

Cosmological Principle The postulate that the universe is, on the large scale, homogeneous and isotropic. The concept of inflation provides an explanation for the observed homogeneity and isotropy.

Cosmology The science of the overall structure and evolution of the universe.

Critical cosmological density Written as ρ_c, this is the critical density separating the 'open' and 'closed' models of the universe. Its value is $\sim 1.6 \cdot 10^{11}$ solar masses per Mpc^3 or -10^{-29} grams per cm^3 for the cosmological concordance model.

Dark baryons Since the hunt for baryons has proved in part fruitless, the hypothesis of dark baryons has arisen: these emit no radiation and are therefore practically undetectable. They may exist in the form of brown dwarfs or as very cold molecular hydrogen.

Dark energy To explain the acceleration in the expansion of the universe, it has been necessary to postulate the existence of a component of its energy/matter content with a 'repulsive' action; cosmologists call it dark energy (because it lends acceleration while not in itself being identified). The cosmological constant is a possible candidate for this energy.

Dark matter Matter probably composed of particles as yet hypothetical, but envisaged by certain extensions of the model of particle physics. Dark matter is the answer to the problem of hidden mass, a problem brought about by the analysis of the dynamics of galaxies and clusters of galaxies and the existence of gravitational arcs (sometimes called 'mirages').

Density Quantity per unit of volume. The terms mass density (ρ), expressed in solar masses per cubic parsec, and luminosity density (ℓ) expressed in solar luminosities per cubic parsec, are, by definition, mass M and luminosity L divided by volume V.

Deuterium Symbol: D. An isotope of hydrogen with a nucleus formed of one proton and one neutron. Easily destroyed in nuclear reactions, the deuterium formed during primordial nucleosynthesis is nowadays difficult to detect.

Digitization Transforming analog data into digital data. In the case of old-style photography, a machine was required to digitize an image, i.e. represent it as a series of numbers on a grid of pixels. Each pixel value is characteristic of the intensity of the signal received at that spot. In digital photography, the image is obtained/coded directly onto the pixels, each of which is an individual photon receptor.

Doppler effect A traveler standing on a railway station platform hears the sound of a train horn drop in frequency as it rushes past. The relative motion of the train to the traveler is the cause of the apparent change in frequency, known as the Doppler Effect. The effect is also observed when we study objects in motion in the universe, for example the rotation of the spiral galaxies. Measurement of the change in frequency (or wavelength) gives an indication of velocity.

Electron An electron is a lepton. It is one of the constituents of an atom, with the nucleons. It is of negative electric charge. Its mass is $9.1 \cdot 10^{-31}$ kg, equivalent to 0.511 MeV.

Electron volt The electron volt (eV) is the unit of energy used in particle physics. It represents the energy acquired by an electron subjected to an electrical potential of one volt. Thus, $1 \text{ eV} = 1.6 \cdot 10^{-19}$ Joules. It is indeed a very small unit (see Appendices).

Ellipticals See Classification of galaxies.

Fermion Particles obeying the Pauli exclusion principle that no two fermions may occupy the same quantum state simultaneously. Electrons and protons are fermions.

Field of view (of a telescope) The area of sky effectively observed through an astronomical instrument. The Moon and the Sun occupy about the same area of the sky (about 30 minutes of arc in diameter), which means that total solar eclipses can occur.

Gamma-rays The most energetic (highest-frequency, shortest wavelength) domain of the electromagnetic spectrum.

Gluon A gluon is a boson that transmits the strong interaction.

Gravitational arcs Distorted and amplified images of distant galaxies (sometimes called 'mirages') due to the presence of a cluster of galaxies along the paths of the photons they emit. In accordance with the laws of gravity, rays of light will be curved by massive objects. Not only are the images of the observed objects distorted, but they are also amplified, appearing brighter due to the concentration of the light rays. Therefore, clusters of galaxies are sometimes known as 'gravitational telescopes'.

Great Debate This revolved around the question of whether or not the observed 'nebulae' were extragalactic. See http://antwrp/gsfc/nasa.gov/diamond_jubilee_debate.html

Hadron From the Greek ἀδρός (*hadrós*, strong). Hadrons are particles that are subject to the strong interaction, unlike leptons. Hadrons are divided into two families: baryons (made of three quarks) and mesons (made of one quark and one anti-quark). Baryons are fermions, while mesons are bosons. The best-known hadrons are protons and neutrons (both baryons), which are components of atomic nuclei.

Heavy water Deuterium oxide, unlike normal water which is hydrogen oxide. It is heavier, since its nucleus contains one neutron as well as a proton, and it is effective in slowing neutrons within nuclear reactors.

Hidden mass See Dark Matter

Homogeneous The same (on average) in all parts.

Irregulars See Classification of galaxies.

Isotope Elements having the same atomic number (the same number of protons) but whose nuclei have a different number of neutrons. Two isotopes therefore have the same atomic number, but different atomic masses.

Isotropic The same (on average) in all directions.

Lepton From the Greek λεπτός (*leptos*, thin, slight). Leptons are particles that are *not* subject to the strong interaction, unlike hadrons. Leptons and quarks are the basic building blocks of matter, i.e., they are seen as 'elementary particles'. There are six leptons in the present structure: the electron, muon, and tau particles and their associated neutrinos.

Large Hadron Collider (LHC) A recently completed particle accelerator which will further our understanding of the nature of matter. It is installed in a circular tunnel 27 kilometers long near Geneva, 100 meters below ground.

Lyman series A series of lines (Lyman-alpha, beta etc.) emitted by hydrogen. A hydrogen nucleus captures a free electron at the first level of its electrical potential, and the surplus energy is liberated in the form of a photon at a wavelength of 1215 Å (Lyman-alpha); the second excitation level corresponds to the Lyman-beta line, and so on... These lines also appear as absorption features when the gas absorbs photons at appropriate frequencies.

MACHO The abbreviation for MAssive Compact Halo Object, as opposed to the WIMP (Weakly Interacting Massive Particle), also a candidate for dark matter.

Mass-luminosity relationship This is obtained from the mass density ρ and the luminosity density ℓ, which are, by definition, mass M and luminosity L divided by volume V. The MLR can therefore be written: MLR $= \rho V/\ell V = \rho/\ell$.

Metals In astrophysics, elements other than hydrogen and helium.

Multiverse Certain models envisage not just one single universe but the possible existence of a very large number of universes (for example, $\sim 10^{500}$!), within each of which the constants of physics could be different from ours, and which would therefore all have different destinies.

Nanometer See Angström.

Neutralino Light particle, possibly the lightest, predicted (though still hypothetical) by extensions of the standard model of particle physics. The neutralino is the favored candidate for dark matter in the universe.

Neutrino The name of this particle means 'small neutral one', drawing a comparison with the neutron, also neutral but of greater mass. The neutrino has been considered either massless or possessing an extremely small mass. Current instruments are as yet unable to detect a cosmological neutrino background dating from the earliest moments of the universe: its detection would constitute a further test of the cosmological model.

Neutron A nucleon, which forms, with protons, the nucleus of an atom. It has no electric charge.

Nucleon A constituent of the atomic nucleus. There are two kinds of nucleon: protons (positive charge) and neutrons (electrically neutral).

Nucleosynthesis (primordial and stellar) The elements of Mendeleev's Periodic Table were formed during two great periods of the thermal history of the universe. Light elements such as hydrogen and helium at the end of three minutes during the period known as primordial nucleosynthesis (Big Bang nucleosynthesis or BBN), and the heavy elements many millions or billions of years later within stars, by stellar nucleosynthesis.

Ω Parameter used in cosmology, corresponding to the density of a component of the matter/energy content of the universe divided by the critical cosmological density ρ_c. The accompanying indicator shows the component in question (e.g. Ω_b corresponds to the contribution by baryons).

Parsec A unit of distance used in astronomy. It is the distance at which the angular separation of the Earth and the Sun is one second of arc. One parsec equals 3.26 light years.

Pauli exclusion principle The principle stating that no two fermions (such as electrons) may occupy the same quantum state simultaneously. This principle helps to explain the electron shell structure of atoms.

Photon A photon (from the Greek *φως* (*phos*, light)) is a boson, and is the particle that transmits electromagnetic interaction.

Pixel A digital image is composed of pixels, a contraction of *pix* ('pictures') and *el* (for 'element'). Each pixel contains a series of numbers which describe its color or intensity. The precision to which a pixel can specify color is called its bit or color depth. The more pixels the image contains, generally the better the resolution, and the more detail it has the ability to describe. A "megapixel" is simply a million pixels. Modern digital cameras contain sensors with 16-21 megapixels.

Planck time Our current knowledge cannot give us any insight into what occurred before the Planck time (10^{-43} s – see Appendices). This involves a blending of General Relativity and quantum mechanics, an objective still to be attained by theoretical physics.

Plasma Gas that is globally neutral, but its electrons are not yet associated with the nuclei (i.e. it is ionized), usually because of very high temperature or exposure to very energetic radiation. Plasma is considered to be the fourth state of matter. The interior of stars consists of plasma at temperatures of many millions of degrees. Before a cosmic age of 300,000 years, the cosmic fluid was also in the form of plasma, and would be so again after the epoch of re-ionization.

Population III Contrary to what the number suggests, this refers to the primordial population of stars.

Processor The 'intelligent' part of the computer system that carries out the instructions of the program. The more processors a computer has, the more operations it can deal with simultaneously.

Quantum number In quantum mechanics, a system is characterized by a family of numbers associated with physical values. For example, the energy of an electron in an atom is characterized by a whole number n. The energy of the electron can vary only by 'jumps': it is quantized.

Quarks Elementary particles (fermions) subject to the strong interaction. Quarks are the constituents of hadrons (e.g. protons and neutrons).

Realization If a process leading to the establishment of a value or set of values is random, i.e. dictated by the laws of chance, this value or set of values is called the realization of the process. This is the case, for example, when tossing a coin. We can apply statistics such as calculating the average value which, for a great number of realizations, will give half heads and half tails, assuming the coin to be without flaws.

Recession of the galaxies Hubble and others showed that all the galaxies they had observed (at the time, a few dozen) were moving away from the Milky Way.

Recombination The process of the formation of hydrogen and helium atoms at a temperature of $T \sim 3{,}000$ K. The so-called 'era of recombination' lies at redshift $z_{rec} \sim 1{,}000$.

Representative sample A set of elements (things, people, data...) selected in such a way that it possesses the same properties as the population from which it is derived. For example, to try to forecast the result of elections, it is necessary to interview a range of people of different ages, sexes, places of residence and social strata.

Rotation curve (of a galaxy) Spiral galaxies rotate. The velocity of any region of a galaxy can be measured by studying the spectral shift (see Doppler Effect) in the wavelength it emits. This velocity can be compared to the distance from the center of the galaxy to establish a rotation curve. Generally, velocity reaches a plateau at a considerable distance from the center. Spiral galaxies do not spin as if they were a solid body (like, for example a spinning top, with velocities proportional to distance from the center), and neither do they mimic the motion of the planets around the Sun, where velocity decreases with distance.

Sloan Digital Sky Survey (SDSS) A comprehensive catalog of galaxies and quasars produced in one of the most ambitious and influential surveys ever realized in the history of observational astrophysics. The SDSS program has so far seen over eight years of observations by one telescope dedicated to the program, with 40 researchers ensuring coordinated data management and analysis. For further information, see http://www.sdss.org.

Spirals See Classification of galaxies.

Stochastic Random, subject to chance.

Supersymmetry (SuSy) An as yet speculative extension of the standard model of particle physics. One of its attractions is that it could predict the existence of numerous (super-) particles which could constitute dark matter.

Surface of last scattering At the time of the recombination, all photons decoupled almost instantaneously from matter and were finally free to propagate through the cosmos. They appear to come from the surface of a sphere for all observers apparently at the centre of this sphere. This is true for wherever in the universe the observer is situated.

Thermal history (of the universe) The description of the evolution of the energy/matter content of the universe, from the Big Bang to the present time. This allows us to define the principal eras of the dominance of one constituent over another, or epochs when changes of state occurred, such as the recombination or reionization of the universe (see Appendices).

Index

CPSIA information can be obtained
at www.ICGtesting.com
Printed in the USA
LVOW02s0824021215
465005LV00009B/138/P

9 781441 988218